STATES OF MATTER,
STATES OF MIND

STATES OF MATTER, STATES OF MIND

Allan Barton

Murdoch University
Perth, Western Australia

with cartoon illustrations by
Andrew Slocombe

Institute of Physics Publishing
Bristol and Philadelphia

ublication may be produced,
itted in any form or by any
copying, recording or
ion of the publisher. Multiple
ith the terms of licences
gency under the terms of its
ice-Chancellors and
Principals.

British Library Cataloguing-in-Publication Data

A catalogue record for this book is available from the British
Library.

ISBN 0 7503 0418 9

Library of Congress Cataloging-in-Publication Data are available

Published by Institute of Physics Publishing, wholly owned by
The Institute of Physics, London

Institute of Physics Publishing, Dirac House, Temple Back,
Bristol BS1 6BE, UK

US Editorial Office: Institute of Physics Publishing, The Public
Ledger Building, Suite 1035, 150 South Independence Mall West,
Philadelphia, PA 19106, USA

Typeset in the UK by Mackreth Media Services,
Hemel Hempstead
Printed in the UK by J W Arrowsmith, Bristol

Dedicated to the ladies in my life

Emma, Megan, June, Rowena and Charmian

PREFACE

There is no 'correct' or 'best' order for presenting the models used to describe matter because the various concepts depend on each other. In the approach I have taken, the first six chapters deal with terminology and fundamental ideas, leading up to thermodynamics and energy. From there on I have considered the 'particles' roughly in order of our familiarity with them. The electron is a good starting point (Chapter 7) because it is the smallest particle with which most of us feel comfortable. I then discuss briefly the other 'fundamental' particles, and protons and neutrons. Chapter 8 deals with atoms, and Chapters 9 to 11 cover molecules, gases and liquids. I introduce solutions in Chapter 12, solids in Chapter 13 and interfaces in Chapter 14. The final chapter takes us from very large molecules to living systems and the Universe as a whole.

Terms likely to be unfamiliar are enclosed in quotation marks. Rather than attempting to give a complete description when each model or concept is first introduced, I refer to it again later when discussing related topics. There is some initial explanation, but your understanding should grow as you meet the new word or idea on subsequent occasions throughout the book. The Index and notes will assist this process.

This book is not a reference work, so the Index has a 'keyword' rather than a hierarchical cross-referenced structure, and is designed to assist browsing rather than to pinpoint particular information.

Notes are included at the end of each chapter. I have used notes to avoid including details in the main text that may not be relevant for everyone. Their main purpose is to provide continuity and connection throughout the book. Many of the notes contain cross references (in bold type) to other sections. With these you can go back to review a term or concept, or forward to explore a new idea from another point of view. Section titles are included (in bold type) in the Index to make this process easier.

The second function of the notes is to suggest further reading. I have chosen sources available from most public libraries: articles in recent issues of periodicals like *Nature, New Scientist, Scientific American, Physics Today, Science, Journal of Chemical Education, Physics World* and *Physics Education*. These are usually more accessible (but less detailed) than monographs or textbooks, and most are less than fifteen years old. Details are included in the Bibliography. This uses the Harvard alphabetical system of referencing. However, I have added the titles of journal articles to give a better idea of the nature of the cited material so you can decide which will be more useful to pursue.

The third purpose of the notes is to acknowledge the sources of some of the ideas I have introduced. Every model or theory or contribution in science depends intricately on the labours and perception of others. The citations allow you to work your way back through this network of ideas.

One can classify the scientific literature as 'primary', 'secondary' or 'tertiary'. Primary literature contains original papers in refereed journals, frequently published by learned societies. Secondary sources include reviews, often in refereed journals or monographs, based on primary articles and citing those sources. Tertiary material generally occurs in the more 'popular' journals, newspapers, magazines or encyclopaedias; it draws on primary and secondary sources but usually does not cite them. *States of Matter, States of Mind* lies somewhere between 'secondary' and 'tertiary': you can find your way back to the original primary publications, but only indirectly, through the secondary sources cited. The notes refer to sources listed in the Bibliography.

Finally, the notes also occasionally include further information and terminology that is interesting and useful but not essential for an understanding of subsequent material.

Allan Barton
Perth, Western Australia, November 1996

CONTENTS

CHAPTER 1

STATES OF MATTER

Some familiar materials like hydrogen are presumably as old as time itself. The 'Big Bang' model for the origin of the Universe proposes hydrogen as the main initial product. On the other hand, some old materials may still appear novel, like the 'fullerenes'[1]†. Fullerenes are 'new' forms of carbon, which have been around in coal deposits much longer than humans have, but chemists and physicists have only just noticed them.

Even more exciting, some materials that chemists and physicists are making have probably never existed in the Universe before. Also, we have access to ultra-low temperatures, much closer to 'absolute zero' than could have existed naturally in the Universe, where the background temperature is three kelvin (minus 270 degrees Celsius)[2]. A temperature of exactly zero kelvin, absolute zero, is unattainable, but with the appropriate technology we can get as close to it as we wish. Microkelvin temperatures (only a millionth of a degree from absolute zero) are routinely available, and we can even achieve temperatures a thousand times closer to absolute zero than this, 'nanokelvin' levels[3].

In trying to answer the questions 'What are things made of, and what holds them together?' we usually classify the numerous materials into various 'states of matter'.

The word 'material' usually suggests something useful or

† Endnotes for this chapter can be found on page 32.

substantial or practical while 'matter' implies a more theoretical and fundamental point of view. However, convention rather than logic determines some of the word usages. We talk about a 'state of matter', not a 'state of material' but say 'materials science' rather than 'matter science'.

How many states of matter are there? Some of us would say just three: 'solids', 'liquids' and 'gases'. Some would include 'plasmas' as a fourth state of matter[4]. Others would want to include 'glasses'. There are also 'liquid crystals', in some ways intermediate between solids and liquids in structure and properties. But rather than try to answer this question now, I shall approach the subject in another way by describing the diverse models that chemists and physicists have proposed for materials, as well as the mental images we use to help us understand them.

1.1 Cake-Making

To appreciate the nature of matter, we must move on from 'concrete operational' thinking (direct observation and experience) to 'formal operational' thinking. In dealing with abstract concepts and things we observe only indirectly through their consequences, we can make good use of analogies and models[5].

It is easier to visualise the structures and properties of new or complex materials if we call on our experience with simpler or more familiar things to provide models. We can improve our insight if we build conceptual models or images representing materials and the ways they behave. Models help us to understand what we observe and to predict what will happen in situations that we cannot observe[6]. Eventually, we may become so convinced of the reliability of a model or the security of an image that we accept it as reality. Then (for us) the distinction between 'model' and 'reality' disappears[7]. While this acceptance of model as reality has advantages, it can also inhibit the acceptance of new ideas and improved models, as illustrated by the persistence of the flat-Earth model. There is nothing wrong with accepting a model as a convenient approximation to reality, but we should do so only until we can develop a more refined or more detailed model[8].

Familiarity with the properties of materials has been essential to

human survival and civilisation. Early technologists successfully managed to achieve a few physical and chemical transformations with very limited conceptual models. These processes mainly involved construction materials, metals, leather, fermentation products, soap, glass and pigments. But recent technological advances with the assistance of the chemical and physical models of the past two centuries have been far more spectacular.

I would like you to imagine your favourite food—it might be a freshly baked cake, or a curry, depending on your taste—and compare the meal itself with the recipe for making it. The recipe is very dull and uninteresting compared with the real thing: it doesn't attempt to describe the aroma, or the flavour, or the texture. However, without a recipe (or previous experience) we cannot reproduce the meal.

The recipe is a kind of a model of the food: if you put certain things together in specified proportions under particular conditions you arrive at the required result. The instructions and list of ingredients in recipes initiate in our minds the thoughts and actions necessary for us to make the real thing. When novices first start cooking, they have to follow the recipe carefully. As they grow more experienced they can take short-cuts and still be successful, and even make improvements to the original. What is more, they can communicate the recipe easily to anyone else familiar with the language they use. Goodman[9] has emphasised the importance of memorising and practising recipes in the process of visualising models. Scientific models are no closer to the reality of materials than recipes are to the reality of food.

Models or images of materials, and of the particles composing them, assist us in visualising complex reality. We can make predictions beyond our existing experience, enabling us to construct useful items, and we can communicate all this information to others. The particular food is unique; the way we write the recipe is not. We can specify quantities in cups or pounds or kilograms, temperatures on the Fahrenheit or Celsius scales or even the scientific kelvin scale, and still get the same result. This is also true of scientific models. There is nothing unique about a particular model, although there is more detail in some models than in others, and some are more useful than others for particular purposes. What they have in common is that they are in some way **idealised** or **simplified**.

1.2 The Greeks Had a Word for It

The first decision to be made in formulating comprehensive models of materials for ourselves is whether matter is continuous and infinitely divisible, or whether in subdividing it we eventually reach a particle that is indivisible. Modern scientists derived the word 'atom' from the Greek *atomos*, 'without cutting'. Democritus taught this around 400 BC, although early Greek scholars themselves disagreed on the existence of 'atoms'. Even today in making complex calculations on the properties of materials there is the choice of using either 'atomistic' or 'continuum' mechanics[10]. Plato and Aristotle also taught symbols for 'elements' and a symmetrical periodic system for the elements, with Plato seeking the ideal in the heavens while Aristotle 'pondered Nature'[11].

In the atomic model, all matter is made up of very small, essentially spherical atoms, usually combined into 'molecules'. An atom contains a minute and extremely dense 'nucleus', made up of particles called 'protons', each with one positive charge, as well as uncharged particles called 'neutrons'. 'Electrons' occupy most of the space of each atom, each electron having one negative charge. The protons and neutrons are of nearly equal mass[12] and the electrons are very much lighter[13]. The abilities of atoms to combine in the form of molecules, which in turn aggregate to form bulk materials, depend mainly on the 'outer' or 'bonding' electrons of atoms. So do the properties of these materials. These electrons are those least strongly held to a nucleus, and the ones most likely to interact with electrons from other atoms to form 'chemical bonds'.

Physicists have shown that protons, neutrons and electrons are not the fundamental elementary constituents of matter as previously believed. Although this is unimportant for some uses of the models (in some aspects of chemistry, for example), a better understanding of materials is possible if we explore more detailed models. The model now widely used in physics incorporates more fundamental matter particles, including 'quarks'[14]. We must also consider how these particles interact. To do this we include in our model the ideas of 'force' and 'interactions', as well as 'force-carrying particles' such as the 'photon', which is the smallest 'particle' of light.

As well as the atom, we also owe to the early Greek modellers the idea of dividing the Universe up into 'matter' and 'forces'. 'Gravitation' acted on two of the four basic elements making up matter in their model, 'earth' and 'water', causing them to sink. The elements 'air' and 'fire' experienced 'levity', the tendency to rise. This distinction between matter and force has clearly been an extremely useful model for succeeding generations of scientists trying to understand how the Universe works. It has been just as important in appreciating the structure of matter as in exploring the galaxies. Although the four elements of the Greeks could combine in various ways, their total quantities were 'conserved' or maintained unchanged, another important idea. (A fifth element, 'ether' or 'quintessence', made up the heavenly bodies.)

As human civilisation has developed, we have modelled matter in ever greater detail and described it with ever smaller particles, satisfying our curiosity and keeping pace with our technology. Technology in turn has provided the means for increasingly detailed experimental probing. We do not abandon our **successful** models entirely, but modify them and continue using them, as appropriate, along with our new models. Many of the **unsuccessful** models have now disappeared without trace, except for those that have become notorious as 'misleading models'. One example is 'phlogiston', an eighteenth century model for combustion before the 'oxidation' model. The phlogiston model proposed that all combustible materials contained this substance, liberating it on heating to leave only ash behind. Another model is 'caloric', invented by Lavoisier as a proposed element that produces heat[15]. Scientists of the day considered phlogiston to be a type of 'earth element' that made a substance flammable. They accepted that they could not isolate phlogiston—they rationalised this on the basis that it would not leave one material unless there was a second material nearby that it could enter. Charcoal was rich in phlogiston, and air could absorb phlogiston during the burning of the charcoal. This is easy to appreciate, as the flickering flames over burning coals do look as though something is 'escaping' from the charcoal. This visual image of burning matter emitting flames may well have reinforced the phlogiston 'habit of mind'[16] and inhibited acceptance of the new oxidation model.

It was noticed that only a fifth of ordinary air had this ability to accept phlogiston from charcoal: scientists then described this gas (now called 'oxygen') as 'fire-air'. The phlogiston model was self-consistent in some ways, and investigations using it led to important chemical facts being discovered. It does not really matter if our model is 'wrong' (inadequate), as long as we are making progress and remaining alert for a better model to appear. It was only when eighteenth-century scientists managed to release oxygen from mercury oxide by heating it that they gradually accepted the new 'oxidation' model. In the burning process, charcoal was not transferring 'phlogiston' to the air, but absorbing and combining with oxygen from the air. They abandoned the phlogiston model, but it had served a purpose.

Those scientists at the forefront of the introduction of new models often believe they have 'discovered' ultimate truths concerning the structure of matter, and support their proposals with religious fervour. Sometimes they suffer persecution in pursuit of their causes. The Greek philosopher Anaxagoras (circa 500 to 428 BC) was the subject of one of the first historically documented cases where a scientist was put on trial for beliefs in conflict with the currently accepted model. His crime was thinking that the stars and planets were solid, physical bodies rather than divine entities[17]. Civilisation owes much to such martyrs.

The ideas of these great thinkers provide us with a succession of models that permit us to visualise what is otherwise beyond our comprehension. Describing them as models in no way diminishes the value of these contributions, or the intellect of those who made them. However, it does liberate us from a feeling of inadequacy because we cannot 'understand' how the world works. In this situation we should be more concerned with 'what' than 'how' or 'why'. We should accept that nobody completely 'understands' the structure of matter in absolute terms[18]. Everyone is building and using increasingly detailed models and images in **attempts** to understand it, and some of us can understand and use models more efficiently or more imaginatively than others. It is noteworthy that innovative scientists like Albert Einstein and Michael Faraday were visual or conceptual thinkers. They made their great advances by manipulating models and developing new ones, and only subsequently translating them into word-models and mathematical models for communication to others. Unfortunately, some of our education systems spend a disproportionate amount of time on superseded models. The result is that students can emerge from the education 'time tunnel' unfamiliar with the models necessary for the 21st century[19].

'Imaging' or 'modelling' is not the only possible approach to the understanding of matter. Another point of view is that all the obvious richness and diversity in material structures, from the largest to the smallest (as well as in living things), are the result of simple 'laws' frequently repeated. On this view our aim should be the 'discovery' of these simple laws. This may be so, but most of us find it preferable to visualise the processes with simplifying models. We find it difficult to express them formally and mathematically in laws and equations, no matter how 'fundamental' these laws may be. Although the **laws** may be simple, it does not follow that **materials** or **systems** must be simple. To describe the infinite variety of trajectories of objects thrown on Earth under the influence of gravitation we need just one 'law' and the particular initial velocity conditions[20].

It is also worth considering what scientists mean by a 'law'. As applied to science, the word carries over from its more general use the suggestion of 'something that must be obeyed'. But in our legal and government systems there are 'unjust laws' still in force, some laws are repealed and even aspects of 'common law' slowly change as society changes. So scientific laws are just particular kinds of model, or ways of summarising models, that we can change as our perspective improves.

1.3 Galloping Horses

Some people criticise science because they say that it explains familiar observations in terms of unfamiliar words and ideas. While this claim has elements of truth, it is much more true to say that **scientific models explain the unfamiliar in terms of the familiar**. There is, in a fictional work[21] about the earliest scientists and technologists, an excellent description of conceptual model-making by the novel's heroes: '...to see the world around them in symbolic form, to extract its essence and reproduce it...'.

Models can be as simple or as complex as we wish, but they are never more than an approximation to reality. We can make models small enough to suit our limited faculties (if reality is large, like the Earth) or conveniently large (if reality is small, like

an atom). We can make models to represent either very long times or very short times, both of which are beyond our comprehension. Models and images are essential to our understanding of materials because we have limited physical abilities to perceive (and therefore comprehend) the world around us. Artists' illustrations of galloping horses painted before the use of photography in the late nineteenth century showed the 'rocking horse' gait with all four legs outstretched simultaneously. (The development of a high-speed photographic shutter originated from the desire to prove that a galloping horse lifts all four hooves off the ground at one point in its stride[22].) If unaided I cannot even perceive a galloping horse correctly, how can I hope to visualise the 'dynamic' or time-dependent structure of water on a molecular scale?

What we observe is always incomplete[23]:

- we can detect (by touch) differences in distances down to only tenths or hundredths of a millimetre

- we can 'see' only in a very narrow band of the electromagnetic spectrum

- we can distinguish separate events no closer than hundredths of a second apart

- the human lifetime is negligibly short on a geological time scale.

One of the aims of scientific models is to make a new or 'unexplained' phenomenon seem 'natural', by analogy with something already accepted as natural from previous experience[24]. Simply inventing or defining a unit of distance or time for very small or very large quantities makes us more comfortable with the ideas. To relate the vast astronomical distances to something familiar astronomers devised the 'light-year', the distance light travels in a year. The eon (originally a long, undefined time but now sometimes used to represent a thousand million years) is another example in that direction. The 'nanosecond' (obtained by dividing a second into a thousand million parts) is useful at the other extreme of short times.

We frequently read that this process or that phenomenon was 'understood' by a particular time, or 'explained' by a certain scientist, or 'discovered' on a particular occasion, but these comments are over-optimistic. Nobody really knows the 'how' or 'why' of the structure of matter. What is meant is that models have been developed which are acceptable because they

- describe the new in terms of the old, or

- express the unfamiliar in terms of the familiar, or

- relate the abstract to the concrete, or

- integrate several previous models in an intuitively satisfying way.

It is even possible to take this idea further, and suggest that what we believe to be objects in the solid world around us are really only images or approximations to reality in our minds[25]. Stephen Hawking said in a 1994 lecture at the University of Cambridge: 'I don't demand that a theory correspond to reality because I don't know what it is'[26]. Because we have such an incomplete physical perception of the nature of matter it would be very bold of us to claim that what we feel and see is the complete and ultimate reality. These claims are even more presumptuous when we come to complex coupled systems, such as the Earth itself[27]. This difference of perception is apparent when we ask people with different educational and cultural backgrounds to describe the same scene or object. If several people look at a tree

- a botanist identifies a *Eucalyptus marginata*

- a forester sees a timber resource

- an ornithologist sees a bird within the tree

- a child notices one fascinating leaf.

Another aspect of the problem of defining reality is the effect of the act of observation on the object being investigated. Although for large-scale or 'macroscopic' objects (those we can see around us) the effect is negligible, this is not true for small particles[28]. It is sometimes useful to ask whether mathematical models describe the behaviour of particles, or whether they describe our **knowledge** of their behaviour, a distinction made by Werner Heisenberg[29].

The scientific process, like most other human activities, requires us to classify and organise information into easily recognisable patterns or models. Whether we realise it or not, we all use models to help us understand how things work. We can have physical models, such as ball-and-stick molecular models, or paper models of chemical compounds like fullerenes[30]. We can have mathematical models like the equation that relates the length of a pendulum to its period, or conceptual models, patterns that help us to organise facts or thoughts in our minds. Models can be static (unchanging) or dynamic (changing with time). Computer models[31] enable us to control the rate at which time flows to assist our perception, and to see things that would otherwise be invisible or obscure, like flows in air and water. The Visualisation Society of Japan has produced an 'Atlas of Visualisation'. This depicts processes as diverse as fluid flow, heat transfer, acoustics and chemical reactions. Of course, this information is even more accessible when presented on a computer monitor[32].

Models may be rigorous and quantitative, or visual and qualitative. A good model is a successful compromise between simplicity and accuracy. In this account of materials, I am emphasising conceptual models, with a little numerical information added in notes, but without any details of mathematical and quantitative models.

It is worth thinking about how far we can afford to simplify reality while still retaining the significance of a model. An old joke[33] tells about a physicist reporting on a programme to develop faster racehorses, 'Let's assume that the horse is a perfectly uniform sphere...'. Our point of view should be that a scientific image, or model, or metaphor, no matter how simple, is useful if it assists any aspect or level of our understanding. However, we must be always aware of its limitations[34].

Whenever we make any form of classification we are using a model, a simplification of reality. We like to classify objects and ideas of all kinds to make them easier to understand and to deal with, but we must always be aware that we are making artificial divisions. We ourselves make the distinction between different classes or categories, not nature[35]. For this reason, no classification or model or theory or concept should be thought of as the last word on a subject. Rather we should just treat it as the newest or most convenient approximation to reality.

'Theories' are also models. The word 'theory' has the same origin as 'theatre', from the Greek meaning 'to view'. Theories and models provide us with a means of viewing reality in a compact, convenient and simplified manner, just as theatre does. Theatre has movement, and we need not only static models of particles and materials and states of matter, but also dynamic models or mechanisms: we need to know how things work. This means that our models of objects (often created independently by different people at different times) are linked by mechanisms. Our faith in the overall model increases if the components fit together and work well. Eventually the composite model may fit our current perception of reality so perfectly that it becomes indistinguishable from reality. We may ask ourselves whether theory ever becomes reality: whether atoms, initially only theoretical and now apparently real, **are** reality[36]? One definition of reality[37] is that a thing is real if it is able to affect a macroscopic physical object. This definition, due to Karl Popper, includes states and products of the human brain, but avoids the quantum behaviour of very small objects[38]. Alternatively, we can define reality in terms of what is changeless, highlighting the importance of 'conservation' principles[39].

1.4 Building a House

Scientific progress is linear rather than cyclic[40], but it is a process of 'successive approximation'. In most areas of human endeavour, we make progress by first making large adjustments, then making progressively smaller incremental changes as we approach our goal. Think about how we go about building a house. First we clear the site, then lay the foundations, and construct the brick or timber walls. Later come the plastering, the painting, fittings, wallpaper and furnishings. A specialist carries out each stage of the work, building on the efforts of previous artisans. If we wish to live in the house at any stage of construction, we can do so, but it becomes more comfortable as it nears completion and we eventually furnish it. The development of a model of matter also proceeds by successive approximation, providing a better fit for our observations as we refine it. Science is 'forever iterating around and towards the unattainable absolute of truth'[41]. From time to time we make larger or smaller changes in our model, but at some time we may decide to construct for ourselves an entirely new model of matter to inhabit[42]. We may retain the old model for other occupiers or other purposes, or demolish it, just as our old house would be retained or demolished.

Whether or not a model is 'real', we must avoid believing in it too completely. We must remain alert to new information or new ideas that allow us to update our model, or even to discard it to make room for a new one. 'Habits of mind' may block our acceptance of new models, just as physical habits or manual operations are difficult to change. There are also examples in science of competing models of reality, coming from different directions and both strenuously defended by their supporters. These often reflect different aspects of a more comprehensive model. The new model is presumably a better approximation to reality, although we can never be sure of that. In trying to understand how the world works, we should freely use as many models as possible, choosing the particular model or combination of models that is most useful on each occasion.

1.5 Hindsight and Foresight

There is a saying that 'Life can only be understood backwards, but has to be lived forwards.' It is easy with hindsight to be critical of our reluctance to discard an old model or accept a new one. Yet it is necessary to understand that this process is extremely difficult and happens rarely[43]. Thomas Kuhn's term for an integrating model is a 'paradigm': a broader model improving on previous models. Different paradigms or models are 'incommensurable':

they do not contradict each other, but they take alternative views of reality. A scientific revolution, analogous to a political revolution, is necessary for a new paradigm[44].

A recent issue of the *Journal of Chemical Education* with the theme 'organising information for the human mind' has collected together several suggestions of models that simplify the process of learning about chemistry[45]. As our collective understanding improves, the models we use need not become more complex—frequently it is more appropriate for them to become simpler.

In our efforts to understand, we sometimes describe a model in a way that suggests foresight or planning ability by a nonhuman system, particularly a more complex biological system. Thus: '*Eucalyptus* trees have developed the ability to survive most fires while producing leaf oils that make the debris around them highly inflammable so that fires ignited by lightning strikes will destroy competing species'. Purists sometimes tell us that this form of expression ('teleology') is inappropriate, even if we are merely using it as a comfortable shorthand for '*Eucalyptus* trees have evolved to ...'. However, we often adopt informal expressions of this kind when using concepts or models. Compare the two expressions

- 'heat flows into the system'
- 'energy is transferred from a source by heating'.

P W Atkins[46] suggests that when we use simplified expressions like 'heat flows into the system' rather than the longer, but more correct, version we can add in a whisper, 'but we know what we really mean'. It does no harm to take liberties with our models, as long as we know what we really mean.

1.6 Keep It Simple

The term used to describe direct, experimental information on the properties of systems (without the benefit of a theoretical basis) is 'empirical'. Sometimes we present this empirical information as

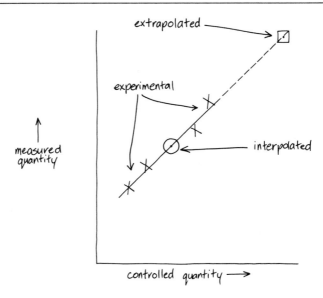

Figure 1.1 Schematic graph illustrating experimental points, interpolated point and extrapolated point.

statements, but on other occasions as simple conceptual models, word-models or mathematical models. If the mathematical equation or scientific model fits the experimental or empirical facts over the range of variables studied, we can obtain information with some confidence **within** this experimental range by interpolation. However, extrapolation **beyond** the range of experimental study provides predictions that are far less certain (figure 1.1).

The use of a more complex relationship or model can allow us to make extrapolations or predictions with more confidence. This is particularly so if the outcomes of the model are consistent with those of other scientific models and with other experimental observations on the same or related systems. Occasionally, if we are lucky, **simpler** relationships arise from the application of a more complex theory or model. Often we use a real system as a model of a less familiar or more complex system. A good example here is our use of the solar system, the Sun and planets held in their relative positions by 'gravitational attraction', as a model for

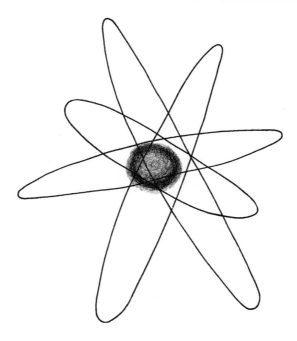

Figure 1.2 The 'solar system' model of an atom.

an atom. In the atom, 'electrostatic' interactions (interactions like those between stationary electric charges) dominate but have the same general form of dependence on distance as gravitational (mass–mass) interactions[47]. Serious scientists used this 'planetary' model of the atom (figure 1.2) as a miniature solar system for only a short time early this century. Nevertheless, it has provided a popular symbol that we almost universally identify with scientific knowledge and progress. Other examples are the use of biological and engineering models for a class of chemical compounds. 'Radiolarians', the protozoans with siliceous skeletons, have structures reminiscent of those of the recently discovered fullerenes.

The simple dome building designed by Buckminster Fuller to distribute stresses uniformly ('geodesically') prompted the name 'buckminsterfullerene' for C_{60}. What is more important, the availability of this life-size model facilitated the proposal, acceptance and popularisation of the novel molecular fullerene structure[48].

Yet another type of approach is to use a model shown to be successful in one application for an alternative purpose. An excellent example is the 'molecular orbital' approach that uses the atom as a model for molecules[49]. Another example is our application of the term 'entropy', introduced as part of the thermodynamic model[50], to the superficially very different situation of loss of information sent by cable or radio. In the presence of constant 'noise' the power needed to transmit information increases dramatically with distance. Also, we use the familiar properties of flowing fluids as models for flows of heat, and of electricity and magnetism. Even the modern quantum mechanical description of superconductors[51] uses terminology based on fluids and fluid dynamics.

1.7 Six Impossible Things Before Breakfast

A good way to start thinking about the particles that make up bulk matter is to ask ourselves a series of questions:

- Do I believe in molecules?
- Do I believe in atoms?
- Do I believe in electrons?
- Do I believe in quarks?

Most of us reading this would answer 'Yes!' to the first three of these questions. But the questions, particularly the fourth, are more like the query 'Do you believe in Father Christmas?' than 'Do you believe the Earth orbits the Sun?' A positive answer really means that the image conjured up by the word 'electron' provides a useful model for thinking about the structure of matter. This is similar to Father Christmas or Santa Claus being a useful model for thinking about the spirit of Christmas.

It is unlikely that we shall replace molecules or atoms as models of materials by anything more useful, but we could not say that about quarks with the same confidence.

Many of us would have as our own conceptual models of the smallest particles in matter something similar to the Greek idea of an atom: dense, solid and indivisible. Without giving up this idea, try exploring an alternative model. Imagine a region of empty space, very small, distorted by some form of extremely intense energy so that it is no longer uniform but concentrated and localised. (We can get an impression of this in two dimensions by pinching and twisting and concentrating the thin rubber of a partially inflated balloon, noting that we have to use energy to cause this distortion.) We now have to accept that by concentrating space or energy like this we have created a particle with properties we describe as 'mass'[52].

Once we can visualise a particle like this, we can assemble the particles we have created to make a structure. As we go on, we should keep reminding ourselves of this idea, of the unity of energy and mass. If you have trouble in believing it, you should follow the example of Lewis Carroll's Alice: 'Why, sometimes I believe as many as six impossible things before breakfast'[53].

1.8 Structure

The word 'structure', when applied to a material, implies an arrangement of particles with some predictability or order, the positions of the particles being 'correlated' or 'coherent'. Crystalline solids or 'crystals' are ordered, coherent, correlated and predictable structures in space (in contrast to gases that are none of these). 'Liquids' have short-range order or short-range structure, but lack the long-range order of crystals. A 'liquid crystal' is a state of matter with long-range order only in some directions, with the order being orientational rather than positional[54].

We do not need to restrict our view of 'structure' to the ordering of things in space or to snapshots at a particular instant. We can generalise the idea of structure to include those arrangements where there is coherence in time, such as in the phenomenon we call 'life'. Ordered activities have structure: we can think of work, unlike heat, as being an ordered form of motion and therefore having structure.

There is a hierarchy of structures making up matter, just as there is a hierarchy of social structures in a community. The structures range from specific one-to-one interactions like two lovers oblivious to all around them, to the much more extensive and impersonal interactions like those in a large crowd of people. In society, our perspective has to change depending on whether we want to know about a particular person, or a couple, or a family or a community. In the same way, we must use different models for materials depending on our particular interest. As the component particles of structures we may consider

- continuous structures such as very large molecules ('polymers') or arrays of atoms ('crystals') which are visible unaided or with a microscope, or

- molecules and atoms, which are too small for us to 'see' directly, or

- electrons and protons, which are even smaller, or

- more 'fundamental' subatomic particles, which are unimaginably small.

Physically observable coherence or structure may arise as a result of ordered processes on a molecular scale, or be the result of random molecular-scale events[55]. Whatever the origin of the structures, studies of them provide clues to our understanding of the states of matter.

1.9 Starting the Jigsaw Puzzle

One of the first things we do in formulating a model is to reduce the scope of the problem to a manageable minimum. We do this by defining a 'system': that particular part of the Universe (usually a very small part) in which we have an interest. As with other words in science, the meaning of the word 'system' is approximately the same as in its everyday use. However, for some purposes we define it more carefully, as I shall show later[56]. The rest of the Universe forms the 'surroundings' of the system, but usually only the **immediate** surroundings are important. The surroundings are separated from the system by a sometimes notional but always carefully defined 'boundary'. We can consider ourselves as part of the surroundings, observing the system but not disturbing it (except in a controlled way if we wish). Systems can be 'real' (open to experimental study) or 'model' (available for conceptual or 'mind' experiments).

The decision on where to draw the boundary is very important, just as in starting a jigsaw puzzle it is helpful if we locate the straight-sided edge pieces before tackling the interior. Living systems also have boundaries, but we can choose the boundary in the most appropriate way. A system can be

- the whole Earth,

- the border of an ecosystem such as a forest,

- the bark of a tree,

- a cell membrane, or

- the effective limit of the electron cloud of a molecule within the cell[57].

1.10 Uneven Progress

Another difficulty we face in exploring the properties and behaviours of materials is that the various aspects of science contain models that are at different stages of development. For this reason we may find it hard to coordinate models in different disciplines. The 'discipline' or area of science we call 'chemistry' uses a basic model of atoms made up of protons, neutrons and electrons, although the area 'particle physics' contains models made up of more fundamental particles. Similarly, the physics of fluid flow processes tends to rely on matter as a continuum rather than to incorporate some of the relevant chemical molecular models. Biologists, while not rejecting particle physics as a model, do not need to make regular use of it. Although these effects tend to be less significant at the leading edges of research, the impression received by those new to science is that of fragmented and even contradictory arrays of ideas.

It is useful to note that the **composition** of the disciplines is also changing. For example, some of the models that were once part of what we choose to call 'chemistry' are now regarded as being in 'molecular biology' or 'biotechnology'. At the same time, new models are arising within 'chemistry'. Models that were previously in 'physics' or 'chemistry' are now part of 'nanoscience'[58]. 'Information science' is tending to expand while 'computer science' is becoming more narrowly defined and competing with 'software engineering'. The result is that as time passes new disciplines arise and grow, some existing disciplines expand, and others shrink. We can imagine the whole of science as a large array of models in a box with movable partitions. Many new models are continually being added, and a few are being removed. At the same time, the partitions in the box are being moved. The relative positions of the models are not affected when this happens, but some of the models finish up in different compartments.

1.11 Whimsicality

Scientists give the impression that they do not have much of a sense of humour, and scientific names, even the 'trivial' chemical ones[59], tend to be uninspired. (One notable exception is where particle physicists use terms such as 'charm' and 'strangeness' in new and imaginative ways.) From our point of view as newcomers to the subject, this has both advantages and disadvantages. On the one hand it is certainly refreshing to meet familiar words in a new subject, and there can be no doubt that these terms are relevant to the behaviour of the models. However, their use suggests that these ideas are frivolous and unimportant, which could not be further from the truth as I shall explain later[60].

In addition, there is a real problem whenever we use ordinary words in an unusual or specialised way. Sometimes we use a common word (with a generally accepted meaning and conveying a particular image) with a specific meaning in conjunction with a particular model[61]. There is no way of avoiding these problems. It will become apparent as we go on that the vocabulary is an integral part of any model so we must ensure that we grasp its precise significance[62].

1.12 Misleading Models

We carry around with us our own images of how things work. Some are helpful, but others hinder our understanding. We become trapped by our own model until we finally manage to escape into a better model in the next 'scientific revolution'. Stephen Hawking[63] pointed out that we had enough information on gravitational behaviour in the late seventeenth century to predict an expanding Universe. Our collective belief in a static Universe was so great that (with the notable exception of Alexander Friedmann) it survived until Edwin Powell Hubble made observations on the red shift in galaxies. These observations were compatible only with an expanding Universe. When reality turns out to differ from what we expect from our internal models, we may describe the result as 'counterintuitive', not consistent with 'common sense'. Of course, in these cases it is our 'common sense' or 'intuition' that is limited.

Images, if relied on too strongly, can inhibit our understanding or appreciation by having a conditioning effect that prevents the acceptance of new ideas. Chemists used to assume that layers of carbon atoms in graphite had to be flat. Now that we know better[64], we see that it is not unreasonable that they should have a tendency to curl up and form closed spherical or cylindrical shells ('nanotubes') of fullerenes (figure 1.3). The flat, two-dimensional diagrams we use to depict molecules distort our view of all carbon compounds, which in practice are dominated by three-dimensional, 'tetrahedral' arrangements.

Although a very successful and useful model, the chemical bond provided us with such a strong image that it delayed our appreciation of 'aromatic' stability. Here some outer or bonding electrons distribute themselves rather evenly between a number of bound atoms rather than being localised in one particular chemical 'bond'. As another example, our belief that certain compounds could not exist inhibited us from looking for them[65].

Sometimes when scientists abandon a model they retain the terminology associated with it, a practice that tends to perpetuate a false model. The use of the name 'organic' for most carbon compounds dates from the time we incorrectly believed that materials associated with living things were fundamentally different from those of inanimate objects.

From some points of view, **all** models of materials are ultimately misleading, because they encourage us to believe that we can visualise molecular and sub-molecular particles in exactly the same way as real life objects. Even more misleading is the very frequent presentation of scientific models without a clear

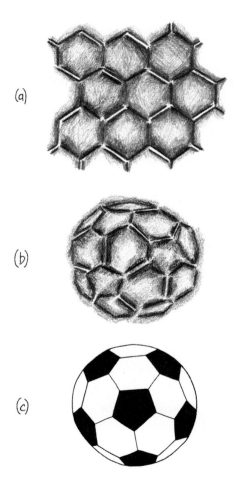

(a)

(b)

(c)

Figure 1.3 (a) A graphite sheet drawn with double and single carbon–carbon bonds. (b) C_{60} drawn with double and single bonds. The resemblance between C_{60} and a soccer ball (c) is evident.

understanding that they **are** models, and our acceptance of them as reality.

1.13 Phases

Associated with the choice of system boundaries is our understanding of what we mean by a 'phase' and what we mean by 'homogeneous'. One definition of a phase is 'a sample of matter with a well-defined set of observable physical properties' and by homogeneous we usually mean 'uniform throughout'. In our physical world we are confident that the systems comprising an iceblock floating in water or sand lying in the bottom of a bucket of water are composed of two separate 'phases', each of which is homogeneous or uniform. If we shake up an oil-water mixture we obtain a cloudy 'emulsion' that is recognisable as a two-phase system under microscopic examination, and which eventually separates into two layers.

- Gardeners use the oil–water emulsion 'white oil' for the control of scale on plants.

- Window-cleaning formulations make use of the two-phase properties of oil–water emulsions.

In a mixture of oil and water with more water than oil, an 'oil-in-water emulsion' forms. If there is more oil, then one obtains a 'water-in-oil emulsion'.

- Milk or cream is an emulsion of oil in water.

- The process of butter-making is the 'denaturing' or breaking down of the protein protecting the fat globules so that an emulsion of water in oil forms: butter.

- In cosmetics, an emulsion of oil in water (cold cream) feels cool because of the evaporation[66] of water whereas an emulsion of water in oil does not.

Each of these (white oil, milk, butter, cold cream) is not a homogeneous, single-phase material, but is 'heterogeneous' (made up of different materials) that can separate into two distinct, homogeneous phases.

Problems arise when we start building scientific models on the strength of our ordinary experience. We are so familiar with 'changes of state' or 'phase transitions', such as ice melting and water vaporising, that we do not realise that these transitions are at all unusual and require explanation. Another problem appears when we look at things on a smaller scale and in more detail. Consider detergent systems. A layer of oriented molecules forms on the surface of the water every time we use soap to wash our hands. Is this a separate 'liquid crystal' phase, even though it is only a few molecules deep? Is the aqueous solution then homogeneous[67]? Does a cell membrane two molecules thick qualify as a 'phase'? Are 'microclusters' of a few tens or hundreds of atoms (with melting points very different from those of bulk materials) separate phases[68]? Why is it that, if we control the temperature and pressure appropriately, we can move materials from a liquid phase through a 'supercritical phase' to a gas phase without undergoing a phase transition?

We realise that we should treat even our well-established concepts and classifications and descriptive terms like 'phase' or 'state of matter' or 'homogeneous' as word-models for our convenience. We should abandon them without regret when their use is misleading.

1.14 How Small Are Atoms?

We can measure the size of a molecule very easily, without any special equipment. Simply pour a little of a 'detergent' (a 'surface-active agent' or 'surfactant') such as oleic acid on to a pond. It accumulates at the surface and spreads as a layer one molecule thick (a monolayer), as demonstrated in an experiment originally performed by Benjamin Franklin[69]. As Lord Rayleigh showed in 1899, the surface tension decreases abruptly at this point, so we can work out the thickness of the film and the size of the molecule.

The length of a surfactant molecule is about 3 millionths of a millimetre. For convenience, we use the 'nanometre' as a unit of length for molecules, so we can say that these molecules are about 3 nanometres long. (As you read on, you will develop a 'feeling' for lengths on a nanometre scale, and become more comfortable about using this new unit.) The nanometre is the diameter of a reasonably large molecule and the term 'nanoscopic' or the prefix 'nano' is now being used as popular shorthand to describe structures of this size. Examples are fullerene 'nanoparticles' or 'nanotubes', biological 'nanofibrils', 'nanocrystals', 'nanotribology', 'nanomachines', 'nanoscience' and 'nanotechnology'[70]. We may call a practitioner of this science a 'nanoscientist' or 'nanoist'. Digital Instruments has registered the trade name NanoScope® for a 'scanning probe' microscope[71]. 'Nanomedicine' could involve devices to attack bacteria and viruses down at their own size level.

Most molecules are smaller than a nanometre in size, so it is convenient, particularly when describing biological materials, to use the broadly descriptive terms

- 'nanoscopic' for nanometre-scale or molecular structures,

- 'microscopic' for micrometre- (or micron-) scale structures, and

- 'macroscopic' for objects visible to us without the aid of instruments.

Scientists usually cope with very large or very small things by expressing quantities as powers of ten, sometimes referred to as 'decade orders of magnitude' or just 'orders of magnitude'[72]. Let us take a length of one millimetre as a length that we can recognise easily and measure fairly accurately with simple equipment. It is a convenient unit for everyday objects. The size of a typical molecule is about six orders of magnitude (six powers of ten) smaller than this, about a nanometre.

We shall first think about the length we reach by starting with a millimetre and going **up** by six orders of magnitude: this is a kilometre, a distance that we can visualise easily. While we clearly need to use different techniques to measure the length of a millimetre on one hand and a kilometre on the other, both are

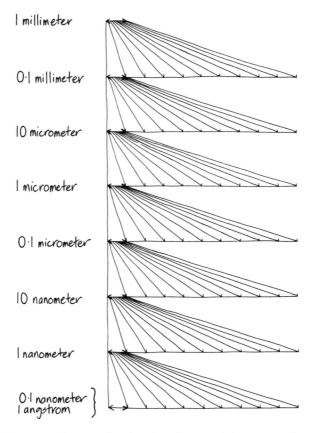

1 millimeter

0·1 millimeter

10 micrometer

1 micrometer

0·1 micrometer

10 nanometer

1 nanometer

0·1 nanometer }
1 angstrom }

Figure 1.4 The seven decade orders of magnitude between a millimetre (which we can measure easily) and an angstrom (which is of atomic size).

within the range of our daily experience. As members of a society regularly travelling by air, we can visualise without difficulty even a further four powers of ten, up to ten thousand kilometres. (This might not be true for an isolated community always restricted by physical barriers, such as on an island or in a mountain valley.)

So if we reverse our perspective in this way we see that six orders of magnitude **less** than a millimetre isn't all that small (figure 1.4)[73]:

(1)   We imagine the millimetre stretched to ten times its length, then divided into ten equal parts. Each division now becomes one-tenth of a millimetre.

(2)   We do this again, so each division is ten micrometres or ten microns. (In a darkened room, a sunlight beam makes dust particles of about this size visible.)

(3)   A further division reaches one micrometre or one micron. (For comparison, the wavelength of visible light is 0.4 to 0.75 micrometre, so the thickness of oil films on water showing coloured interference patterns[74] is of the order of a micrometre. This is about the resolution of an optical microscope, and corresponds to a few thousand atoms across.)

(4)   When we expand and subdivide the micrometre ten times, each division is one-tenth of a micrometre.

(5)   Another operation like this provides a unit length of ten nanometres, the size of a large molecule.

(6)   One more division brings us to one nanometre, 10^{-9} metre. This is the most convenient unit for chemical structures (molecules).

(7)   A final subdivision provides the radius of a typical atom: one-tenth of a nanometre, or one 'angstrom'. Chemists often use this unit for chemical bond lengths.

When we are talking about materials,

- the nanometre is a convenient unit for chemical structures (molecules),

- the micrometre is suitable for single-phase (chemically homogeneous) particles or films, and

- the millimetre is usually appropriate for fabricated (multi-phase) objects.

Although atomic dimensions are very different from macroscopic or human-scale dimensions, we should not be afraid of them: they are conceptually attainable and (with instrumental aids) physically observable and measurable.

For **times** at the molecular scale it is convenient to go down to much smaller units, beyond the nanosecond, to the picosecond (one-thousandth of a nanosecond) and the femtosecond (one millionth of a nanosecond). So the nanometre is the characteristic length for molecules, and the characteristic times for molecular processes tend to be intervals ranging down to the femtosecond.

We may still have trouble with the very large **number** of atoms in an ordinary object. We cannot really appreciate the multi-billion number of people on Earth, let alone the far greater number of molecules in a glass of water. However, these large numbers do have the advantage that we can get accurate information on the particle populations by taking average or mean values. In the same way, we report from census results various types of average properties and their distributions[75].

It is perhaps unfortunate that most books starting out to describe the composition of matter make a point of stressing how **small** atoms are, overselling their smallness. Until recently even many scientists believed that we would never be able to 'see' individual atoms. Now we can do this by ingenious techniques[76], and even move them around one by one in a controlled way. We are also able to use 'artificial atoms' as models of the real thing[77]. Steven Chu[78] has pointed out that when we turn a page the distance between each finger and the paper is of the order of atomic dimensions. It is the redistribution of charge resulting from the interaction between the electrons on our skin and those on the paper that allows us to grip the paper and turn the page.

Endnotes

[1]See **9.14 Fullerenes**.

[2]The kelvin temperature scale has units of the same size as the Celsius or centigrade degree scale but has zero at the absolute zero rather than the freezing point of water.

[3]See **5.3 Absolute Zero**. Some systems can be considered to have **negative** temperatures **if temperature is defined in terms of the Boltzmann equation** and if populations of energy levels are manipulated to create a population inversion, with the highest energy level the most populated. This can be discussed in connection with nuclear spin distributions, and even here a temperature of exactly zero is unattainable. For further reading see Owers-Bradley (1993).

[4]See **8.11 Plasmas**.

[5]Fortman (1993).

[6]Model-building is associated with those aspects and techniques of learning described as 'constructivism': acquiring knowledge by constructing it for ourselves rather than learning by rote when we are taught in the 'transmission' mode. Further reading (with an emphasis on concept mapping) Ebenezer (1992).

[7]Further reading Davies (1985, Chapter 4).

[8]Leatherdale (1974, p 42) identifies a set of four distinct meanings for the word 'model' as well as for other terms such as 'analogy' and 'metaphor' in science, but here the words 'model' and 'image' are used with the broadest possible ranges of meanings.

[9]Goodman (1990).

[10]Kolb (1977). Discussed further in **10.4 Computer Simulation**.

[11]See **8.6 Periodic Elements**. Further reading von Baeyer (1993), Pohl (1993), Heilbronner and Dunitz (1993), Hoffmann (1993).

[12]Of the order of 1.6×10^{-24} gram. For the distinction between 'mass' and 'weight' and further discussion on mass see **3.3 Newton's Model**.

[13]Of the order of 10^{-27} gram.

[14]See **7.8 Colourful Quarks**.

[15]Further reading Vaughan (1994), Sparberg (1996). For 'caloric' see also **5.2 Thermodynamic Model**.

[16]Holderness (1994).

[17]Arnau *et al* (1995).

[18]Feynman (1985, Chapter 1), Harrison (1993).

[19]Dobson (1993).

[20]Further reading Davies (1985, Chapter 4). 'Velocity' is speed in a particular direction.

[21]Auel (1991).

[22]Further reading Graham (1991), Zewail (1990).

[23]This is discussed further in **2.2 Perspective**.

[24]Leatherdale (1974, p 56).

[25]Further reading Postle (1976, Chapter 3), Stapp (1982).

[26]Hawking and Penrose (1996).

[27]Discussed in **15.12 Life**.

[28]Discussed in **4.4 Heisenberg May Have Been Here!**. Further reading Davies (1980, Chapter 6).

[29]Further reading Davies (1980, Chapter 7).

[30]Beaton (1992). **9.14 Fullerenes**.

[31]See **10.4 Computer Simulation**.

[32]Further reading Earnshaw (1993).

[33]Quoted by Graham (1991).

[34]Further reading *New Scientist* (1993).

[35]Further reading von Baeyer (1993).

[36]Wynn (1992).

[37]Rae (1986, p 65).

[38]See **4.1 Ultraviolet Catastrophe**.

[39]Further reading Postle (1976, Chapter 11), Stapp (1982).

[40]Further reading Horgan (1996). Periodic table: see **8.6 Periodic Elements**.

[41]Lovelock (1988, p 12).

[42]See next section and **7.8 Colourful Quarks**.

[43]Further reading Sparberg (1996).

[44]Kuhn (1970), Gleick (1988, p 36), Holderness (1994). Thomas Kuhn died in 1996 aged 73. David Hull (1996) concluded a commentary on Kuhn's contribution to science with: 'It would be a great shame if Kuhn were remembered only for his claim that different paradigms are incommensurable with each other'.

[45]*Journal of Chemical Education* (1994). See Chapter 9, particularly **9.5 Classifying Matter**, **9.12 Handedness** and **9.9 Carbon Compounds**, for details.

[46]Atkins (1984, p 24).

[47]See **8.1 Atom Models**.

[48]See **9.14 Fullerenes**. Further reading Curl and Smalley (1991), Curl (1992), *Journal of Chemical Education* (1992).

[49]See **9.2 Molecules** and **9.13 Aromatic Model**.

[50]See **5.2 Thermodynamic Model**.

[51]See **13.7 Superconductors**.

[52]Further reading Postle (1976, Chapter 4).

[53]Lewis Carroll's character and scenarios are popular with particle physicists. See also **4.1 Ultraviolet Catastrophe**.

[54]See **15.8 Liquid Crystals**. Further reading Collings (1990).

[55]Discussed in **3.8 Fractals**. Further reading Atkins (1984, p 179, Chapter 9, 'Patterns of Chaos').

[56]See **2.1 Scientific Jargon** and **5.2 Thermodynamic Model**.

[57]Further reading Lovelock (1988).

[58]See **1.14 How Small Are Atoms?**.

[59]Smith (1992).

[60]Further reading Farmelo (1992).

[61]Feynman (1985, Chapter 1).

[62]The 'Deborah number' (**11.4 Viscous Flow**) and the 'quark' (**7.8 Colourful Quarks**) are rare examples where literary allusions are light-hearted without introducing confusion or irritation. See also **2.1 Scientific Jargon.**

[63]Hawking (1988, p 42).

[64]Further reading Curl and Smalley (1991).

[65]See **9.3 Bond Models**, **9.9 Carbon Compounds**, **9.13 Aromatic Model**, **9.14 Fullerenes** and **9.11 Nonexistent Compounds**.

[66]See **5.8 Phase Changes**.

[67]See **14.7 Surfactants** and **15.8 Liquid Crystals**.

[68]See **9.10 Microclusters**.

[69]de Gennes (1992).

[70]Discussed in **9.10 Microclusters** and **14.14 Nanotribology**. Further reading Langreth (1993), Stevenson (1996), Stix (1996), Bethell and Schiffrin (1996).

[71]Described in **6.2 Probing Matter** and **8.10 Surface Images**.

[72]Atkins (1984) deals with powers of ten in temperature in this way, and there is a related publication 'Powers of Ten' concerning distance.

[73]Further reading on powers of ten: Kaye (1993, Chapter 1). Weber (1992, p 349) quotes some alternative amusing decimal prefixes, including 10^{-12} boos = 1

picoboo, 10^{-9} goats = 1 nanogoat.

[74]Described in **4.2 Wave States**.

[75]This is the great strength of the thermodynamic model, **5.2 Thermodynamic Model**.

[76]See **8.10 Surface Images**.

[77]Kastner (1993). See also **7.4 Electron in a Box**.

[78]Chu (1992).

CHAPTER 2

LANGUAGE OF MATTER

As well as surmounting the barriers built up by our misleading models, we also have to overcome the problem of the special language developed by scientific modellers. We do this by treating these specialised terms as word-models, with built-in aids to understanding, so they become an integral part of our own conceptual models. Before we can use them, we must overcome our distrust of new words and new ideas.

2.1 Scientific Jargon

We look at a physics journal and see words like 'quanta', 'hadrons', 'quarks' and 'leptons', to say nothing of 'charm', 'strangeness' and 'colour'[1]†. Why are all these names necessary? The same thing occurs in all fields of science[2]. Surely we can talk about material in everyday terms rather than jargon that seems to introduce barriers to communication? Unfortunately, we cannot avoid the initial effort of learning new terms. We cannot separate the models from the words we use to describe them. All terms describing observations carry theoretical images about models as well as experimental information[3]. Sometimes we use a word-model to get across a difficult idea or concept just as we use a mathematical model for the same purpose.

We cannot avoid new words when describing scientific models. At

† Endnotes for this chapter can be found on page 52.

one time these new words or 'neologisms' tended to have Greek or Latin origins but scientists are now devising words like 'gluon' to describe particles holding matter together. Such brand-new words are often more helpful to us in our model-building than a common word given a special definition, because an established meaning or implication of an existing word may be hard to escape.

An additional difficulty is that, for historical reasons, areas of study and their associated models occupy separate scientific disciplines. 'Physics' deals with particles making up atoms, while 'chemistry' takes over for the interaction of atoms, ions and electrons into assemblies of molecules and some of the molecular properties of materials. Then 'physics' returns to deal with large-scale properties of materials. The activities of 'physical chemistry' are similar to but quite distinct from those of 'chemical physics'. From the outside, these discipline boundaries are not necessarily very logical, but people working in the disciplines usually obey quite strictly the conventional divisions (described as 'tribal rules' by James Lovelock)[4]. There are sub-disciplines, with physics and chemistry being subdivided into subjects such as 'particle physics' and 'organic chemistry'. From time to time entirely new subjects develop, such as 'materials science' or 'geophysiology'[5]. Scientists in one discipline sometimes use as tools the models of other disciplines, models they may not fully understand[6]. Sometimes the terminology changes as we change perspective. Words may even be used with different meanings in different models. To the chemist, interactions between atoms can be 'strong' or 'weak'. They can be as 'strong' as those chemists describe in terms of a chemical bond. They can be as 'weak' as those between 'colloidal' particles and similar materials described as 'complex fluids' or 'structured fluids'[7]. However, to the physicist thinking about 'weak interactions' and 'strong interactions', the terms 'weak' and 'strong' conjure up very different images[8].

While abbreviations and acronyms are acceptable within a discipline, sometimes specialists overdo them when they are trying to communicate with those outside the discipline. Occasionally, a set of abbreviations or an acronym is so appropriate and useful that it quickly enters our language as a fully accepted word. Few people who now use the noun 'laser' and even the verb to 'lase' realise that it originally meant **L**ight **A**mplification by **S**timulated **E**mission of **R**adiation[9]. Generally, however, the use of such abbreviations imposes barriers to communication not only within science but also between scientists and the general community. We all like to use shorthand if we can, and terms used by professionals when picked up by the media and used inappropriately become misleading. An example is 'quantum leap', a term now using the word 'quantum' to imply a **great** change rather than emphasising the original meaning of discontinuity[10]. The recent use of the prefix 'nano' meaning 'of molecular dimensions' may be appropriate if the dimensions are of the order of a nanometre (one-millionth of a millimetre). However, we are likely to find this term being used less precisely, just as we over-use the prefix 'micro'[11]. Logically, we should use the term 'microscopic' only to mean 'of the order of a micrometre or micron' (one-thousandth of a millimetre), but until recently it was used to refer to anything too small to see unaided. It is now convenient to use the newly invented word 'nanoscopic' for molecular dimensions, of the order of a nanometre.

The terminology or jargon of an area of science reflects its models in a shorthand form, and we cannot get any real understanding of a model without acquiring familiarity with its language. Although at first it might appear awkward and unnecessary, the jargon is an integral part of any model and we cannot avoid it completely: 'to name is to know'[12].

2.2 Perspective

From a human perspective there are tremendous ranges in the sizes of quantities to comprehend. Time is an important dimension in materials and processes. Significant times in the study of matter range from thousands of millions of years down to less than a millionth of a nanosecond (a femtosecond). Distances cover as wide a range, even for one material. As an example, I suggest you think about the various models we can use in discussing seawater. The water in the oceans has:

- structured movements (currents and tides) spanning thousands of kilometres (millions of metres),
- waves of the order of metres in height,
- molecular and atomic features on the nanometre scale, and
- subatomic structures that are almost unimaginably small.

Looking at another example, our macroscopic observation of gas pressure is a result of a very large number of relatively small molecular interactions.

One thing that makes it a little easier to cope with the great range of dimensions is that events or objects characterised by great disparity of size tend to have little effect on each other. Atomic interactions within a water molecule and molecular interactions between water molecules are just the same whether they are in a laboratory test-tube or in an ocean. Further, from the point of view of currents and tides we can treat the oceans as if they are continuous liquids, and completely ignore the existence of molecules. We can isolate a limited range of length scales, or times, or masses when we are considering the properties of materials or modelling a particular set of properties. It is also useful to recognise those relationships which are true regardless of the scale on which we measure them[13].

In nature we often observe processes that at first change very slowly, then accelerate more and more steeply until interrupted by other factors. The growth of a human population before resource limits are reached is a good example. We can use a mathematical exponential function to describe this process, and the mathematical expression 'exponential growth' has entered our everyday language for such situations. In exponential growth, a quantity that doubles after a certain time interval doubles again in the next time interval of the same length, then doubles again in the next equal time interval. The inverse function of an exponent is a natural logarithm. If we take the data showing an exponential rise (such as population growth) and plot them as a logarithmic function we obtain a straight line (figure 2.1).

We can show with some simple mathematics that whenever there is a situation where the rate of increase of a quantity is proportional to the amount already there, an exponential increase occurs. For example, in modelling the industrial revolution, it is reasonable to use a simple model in which the rate of production of steel by a society is proportional to the amount of steel already in use. This simple model results in an exponential rise. Similarly, the assumption that birth rate is directly dependent on the population results in exponential population growth (until another factor intervenes). Relationships in Nature often take the form of an exponential rise interrupted by an 'exponential decay'[14].

If the likelihood of an event depends on the exponential function

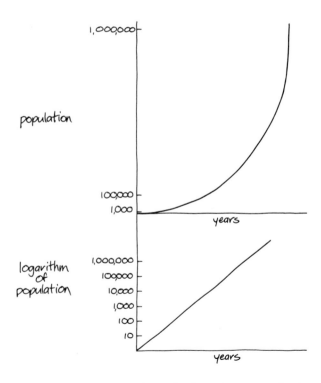

Figure 2.1 Top: exponential growth of population as a function of time. Bottom: the same information on a logarithmic scale generates a straight line relationship.

of a quantity, then the likelihood of the event will follow a 'logarithmic' distribution rather than linear distribution. (In science and nature, the rates of most processes do tend to have an exponential variation with temperature, for example.) We are programmed to think of a 'linear' flow of events as being 'natural'. We count each day one by one in a linear fashion as it follows the previous one. In a similar way we started to count by checking things off on our fingers, and measured lengths by the number of our paces. However, in many situations the most 'natural' distribution is an exponential one, so that it is only by using a logarithmic scale that we obtain the linear distribution we find so satisfying. (The 'decibel' scale in acoustics, the hardness scales for materials and the energy scales for earthquakes are examples.)

We can characterise the nature of the curvature of an exponential function of a quantity changing with time (that is, the amount of

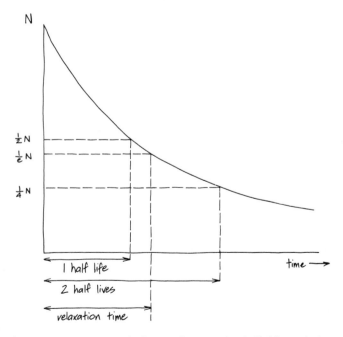

Figure 2.2 Exponential decay, showing the half life and the relaxation time.

acceleration it represents) by the 'relaxation time' or 'half-life' (figure 2.2). (The half-life and relaxation time are numerically similar but not identical. It is customary to use half-lives to discuss nuclear decay processes and relaxation times when describing most other physicochemical processes.)

Systems and processes have this exponential time dependence with a characteristic time acting as a boundary between two regimes. For times much shorter than the relaxation time they act in one way, and for times much longer than the relaxation time they act in another. Only for interactions with a 'time scale' of the same order as the relaxation time can we obtain details of the dynamic processes. Dynamic processes are time-dependent processes where a system under stress relaxes to an 'equilibrium' state. The typical frequency of a process is the reciprocal of its relaxation time. ('Relaxation' is another of those 'ordinary' words to which scientists have given a specialised meaning for a particular purpose. I discuss the precise meanings of 'state' and 'equilibrium' later[15], but at the moment you can accept their 'ordinary' meanings.)

Although our aim is usually to identify the correct time scale or distance scale for a problem, we must also be able to investigate problems that have no characteristic scale. Under certain conditions, correlation lengths diverge so that there is no 'characteristic length', a condition of 'criticality'. There are also transport and relaxation processes where there are no unique relaxation times, where the average 'waiting times' for events diverge, particularly in disordered materials such as polymer films and glasses. The mathematical model of 'stretched exponential functions' has been introduced for some of these situations. Again in our use of models we have to be cautious about applying any particular model (in this case exponential change and characteristic relaxation times) indiscriminately to systems where it is inappropriate[16].

2.3 How Long Is a Piece of String?

Think about all the pieces of string in the world. Although it is possible to have pieces of string that are extremely long (kilometres in length during manufacture, for example) or very short (when we trim off the excess after tying a knot), a 'typical' piece of string is about a metre long[17]. This is because our world is full of objects about a metre in circumference that we need to package. Very long or very short pieces of string are of no particular use to us.

In much the same way, a gas has a typical distance of free travel between collisions, the 'mean free path'[18]. This distance is intermediate between the shortest possible (a molecular diameter) and the longest possible (the dimensions of the container). Gas molecules separated by distances very much **greater** than the mean free path are effectively independent of each other. Typically in a gas there are few neighbours at distances very much **less** than the mean free path. Consequently, the overall properties of gases (unlike those of condensed matter, solids and liquids) are largely independent of the natures of the constituent molecules or atoms and determined by the mean free path. The value of such a 'typical' quantity is therefore very important in our understanding of matter and its interactions.

2.4 Condition Critical

As noted above, there are some phenomena where events over a wide spread of length scales are all significant, at the 'critical points'[19]. Critical points exist in phenomena of all kinds, for example the shape of a granular pile, such as a pile of sand[20]. As we pour dry sand onto a growing pile, the grains organise themselves. The movement of each grain influences others around it, until the characteristic and reproducible angle of slope of the pile is achieved. This combination of 'self-organisation' and 'criticality' we call 'self-organised criticality'[21]. It combines

- a sensitivity to small events (such as one grain of sand dislodging many others), with

- a robustness in the nature of the final state (such as the final shape of a sand pile).

There is an insensitivity to lengths and times. No separation in distance or time is sufficient to make one grain of sand immune to the behaviour of another[22]. It is interesting that if the sand contains foreign bodies like small pebbles, disorder[23] is introduced into the regular packing of sand grains limiting the extent of the criticality.

There is in self-organised criticality a feedback mechanism, with information from existing conditions influencing future ones. This ensures a type of 'steady state'[24]. The apparently simple process of the formation of a sand-pile is fundamentally complex, but it can provide valuable information on systems of similar type that are even more complex. We can consider a process at several levels or from different points of view. Relative to one sand grain an avalanche of grains is a very complex, composite event. Looking at the sloping sides of the pile as a whole reveals a return to a simpler, more understandable behaviour.

If a phenomenon conforms to the self-organised criticality model it should exhibit particular kinds of pattern in the apparently random fluctuations. One is the occurrence of string-like sequences in the locations of events, these sequences appearing to be similar on all scales as we successively increase the magnification. We call these sequences 'fractals'. If the grains adhere to each other, the result is the formation of 'aggregates' with a fractal structure[25].

A related phenomenon is the observation that mixtures of particles of different sizes become segregated by shaking (for example, pebbles 'floating' to the surface of sand when shaken in a bucket). Computer simulation modelling suggests that here also there is a critical condition. In this case it is the ratio of diameters of the particles that is critical, separation occurring if the diameter of the larger particles is more than about 2.8 times that of the smaller particles[26].

Critical phenomena also occur in materials where there are discontinuous transitions between phases as a result of change of temperature or (in the case of mixtures) change in composition. There are pressures and temperatures where the distinction between liquid and gaseous phases disappears. There are temperatures, pressures and compositions where miscible liquids become immiscible and there are temperatures where magnetic materials become nonmagnetic. We observe critical phenomena also in surfactants, liquid crystals and gels, as well as in phenomena like turbulence. All these systems have in common a range of fluctuations that range in size from molecular up to those physically observable on a 'human' or macroscopic scale[27].

2.5 Smaller and Smaller Probes

We can take advantage of the great diversity of scales by probing small particles with even smaller particles. The experiment

proposed by Ernest Rutherford in 1910 used alpha particles
(positively charged helium nuclei) to bombard gold atoms. In this
experiment the scattering showed that atoms have very small,
charged nuclei[28]. More recently in a similar way, high-energy
electrons have been used to probe the nucleus itself. Here the
nature of the scattering shows that protons and neutrons have
charged, point-like constituents ('quarks')[29].

An example of the reverse kind of situation—large particles
demonstrating properties of smaller particles—is 'brownian
motion'. We can see macroscopic particles in a fluid[30] (such as
colloidal particles suspended in a liquid, or smoke particles in the
air) undergoing erratic movements as a result of molecular
impacts. (The original microscope observation by Brown in the
early nineteenth century was of pollen grains suspended in water.)
Red blood cells have interfaces with the blood fluid which are free
of surface tension effects and very flexible. They can be seen to
undergo shape changes caused by random impacts from adjacent
molecules[31].

Particles are not the only kind of probe. The development of laser
light has provided us with methods of not only studying small
particles, but also controlling their movement[32].

2.6 Limits and Dimensions

The previous section's reference to the use of smaller and smaller
probes reminds us that we must be aware of the limits of such
processes. Models often describe 'limiting' behaviour, behaviour
which is not physically accessible. Examples are sources of
electrical charge so physically small that we can treat them as
'point' charges, and electrical capacitors with parallel plates of
'infinite' dimensions. Limiting models are easier to visualise and
more convenient to describe mathematically. The results obtained
from these models also have a satisfying degree of symmetry, as
discussed below. We can devise **real** experiments that
approximate to these models. We do this by having very small
sources of charge, or capacitor plates which are very large

compared with their distance apart. However, they are never exactly the same as our limiting **models**[33].

As well as its common meaning of 'certain specified size', the word 'dimension' has particular scientific and mathematical uses. One of these refers to the number of coordinates, as in 'three-dimensional' space, and one of the unhelpful models we carry around with us is the uniqueness of three-dimensional space. This concept is deeply ingrained not only because of our physical experience but also because (in the words of David Ruelle[34]) it has been 'hard-wired into our brains' by evolution. We may find it difficult to accept models of higher dimensions, although most mathematicians and many physical scientists learn to use them routinely[35]. Images of dimension **smaller** than three pose no problems for us. We can visualise a two-dimensional curtain, a one-dimensional straight wire, and even a zero-dimensional point as portions of everyday, three-dimensional space[36].

Dimensions are also important in the mathematical analysis of physical phenomena, because we can describe all physical quantities as various products of just three: mass, length and time[37]. We call this procedure 'dimensional analysis'.

What we often do not appreciate, however, is that the **effective** dimension of an object depends on the degree of resolution, as outlined by Benoit Mandelbrot. If we look at a ball of string from a large distance, it appears as a zero-dimensional point. As we approach more closely, we see a three-dimensional figure, and even closer an array of one-dimensional threads appears. On still more detailed examination each thread becomes a three-dimensional column, and so on, as we get down to the fibres making up each thread. Ultimately in principle we could see zero-dimensional atoms[38].

To really appreciate some of the models of materials we must free ourselves from our three-dimensional world. We can put ourselves in the position of beings living in an environment of fewer than three dimensions, and imagine them trying to visualise an 'unnatural' three-dimensional Universe. It is as though we are two-dimensional creatures living in a two-dimensional world, trying to describe the surface of a sphere with a flat-plane model. We then realise that we can describe it more easily and 'naturally' in three dimensions.

Probably the most widely used higher-dimension concept is that of 'spacetime'. This arose in the imagination of Henri Poincaré and Heinrich Lorentz at the end of the nineteenth century when investigating the symmetry in James Clerk Maxwell's electromagnetic equations[39]. We can regard a dimension such as length or a time duration as a projection of spacetime on to space, or on to time[40]. Stephen Hawking[41] explained why it is possible that life as we know it can exist only in those regions of spacetime where one time dimension and three space dimensions are not 'curled up small'. This is a reasonable explanation, but there are other good reasons for us to develop our imagination for perceiving higher dimensions. We can model the structures of solids without long-range order, such as glasses, as projections onto three-dimensional space of structures that are regular in higher dimensions[42]. There are many situations where the effort of familiarising ourselves with such scientific images is worthwhile.

Using the concept of dimensions even more generally, we can treat each independent variable in a problem as a dimension. (This applies to real life problems as well as scientific ones.) The introduction of dimensional complexity requires us to ask whether a particular problem can be solved at all, and it is convenient to classify problems as

- undecidable (we can 'prove' that some theorems can never be proved!), or
- noncomputable, or
- intractable.

Mathematical solutions are available in some situations that would otherwise be intractable, as long as we accept an average result rather than a specific result. Other problems may be noncomputable[43].

2.7 Microgravity

In developing models for ourselves free of conventional constraints we must not overlook gravity. On Earth there is an almost uniform gravitational background. Although we can generate **greater** 'artificial' gravity using a centrifuge, only recently with regular space shuttle flights have we been able to carry out processes under **lower** gravity. These 'microgravity' conditions enable us to grow crystals from solutions or melts undisturbed by convective processes caused by density differences. They prevent distortion of crystals such as proteins involving very weak intermolecular interactions. They also enable observation of interfacial effects such as 'thermocapillary' flows, which gravitational effects obscure on Earth[44].

2.8 Symmetry

Unlike many of the images that we require in modelling, symmetry is not only familiar to us in everyday life but it is also the basis of pleasurable sensations, both visually and in music[45]. The drawings of M C Escher show the strong link between art and symmetry[46]. Symmetry is also a source of professional satisfaction to most scientists. As George S Hammond said when referring to fullerenes, 'chemists have an instinctive delight in symmetry'. Others have made similar comments on the aesthetics of fullerene structures[47].

The obvious symmetry in crystalline solids is a result of the order in the arrangement of their molecular particles (atoms or molecules)[48]. The form of symmetry in fractals results from 'aggregation' by random, diffusion-limited molecular events[49]. If we see what appears to be a crystal with fivefold symmetry, experience tells us that it cannot be a true crystal. Units with fivefold symmetry such as dodecahedra cannot be close-packed to fill space, and materials showing fivefold symmetry are not real crystals but 'quasicrystals'.

The existence of symmetry implies that during some operation or process where other things change a particular quantity is unchanged or conserved. 'Spin' is an essential aspect of the fundamental particles that we can visualise most readily by means of symmetry, the symmetry of a dinner plate or a playing card[50].

The absence of symmetry in the fundamental particles disturbs some scientists. The 'chirality' or 'handedness' of molecules and of crystals reflects this asymmetry. Nature not only lacks symmetry in materials but also in flows of energy and time. Energy flows from hot to cold, not the reverse, and time flows in only one direction[51].

However, to use these ideas fully we have to modify our view of what 'symmetry' means. We can extend the idea of symmetry beyond the familiar geometrical examples. The energy change involved in lifting a mass depends on the height difference but not on the absolute height above ground level, an example of so-called 'gauge symmetry'. In this example there is symmetry associated with the choice of the reference height[52].

Endnotes

[1]See **7.8 Colourful Quarks**.

[2]Further reading Farmelo (1992).

[3]This is an important aspect of Thomas Kuhn's philosophy of science. See **1.5 Hindsight and Foresight**.

[4]Lovelock (1988, p xiv).

[5]**15.12 Life**.

[6]Further reading Ruelle (1991).

[7]Discussed in **9.3 Bond Models**, and **15.1 Soft Matter**. For colloids, see **14.7 Surfactants** and **15.6 Colloids**. The models for interactions **within** atoms, the structures of atoms, are described in Chapter 7.

[8]See **6.1 Interactions** and **7.1 Beyond the Atom**. Another example (the different images associated with the word 'crystal') is provided in **15.6 Colloids**.

[9]See **4.2 Wave States**.

[10]See **4.1 Ultraviolet Catastrophe**.

[11]Further reading Ball and Garwin (1992).

[12]The use of specialised terminology (particularly abbreviations) has been kept to a minimum here, but those terms which are introduced are essential for an appreciation of the scientific models. These terms are explained not only when they are first introduced, but also again when they are re-introduced or used in discussion, so that they gradually become familiar, part of our integrated model describing the structure of materials.

[13]See **3.9 Scale**.

[14]This is illustrated in the Daisyworld models of whole-Earth relationships (**15.12 Life**). Further reading Lovelock (1988), Ruelle (1991).

[15]Discussed in **3.1 States** and **5.2 Thermodynamic Model**. The 'Deborah number', the ratio of relaxation time to observation time scale, is described in **11.4 Viscous Flow**.

[16]Further reading Scher *et al* (1991), Wilson (1979). See also **2.4 Condition Critical.**

[17]Mehta and Barker (1991).

[18]The mean free path is the average distance which particles in gases travel between consecutive collisions: see **10.5 Ideal Gases**.

[19]See **12.4 Phase Diagrams**.

[20]See **11.7 Granular Matter**.

[21]Further reading Bak and Paczuski (1993).

[22]Further reading Cardy (1993).

[23]See **3.7 Chaos** and **3.10 Dirty Systems**.

[24]See **3.2 Dynamics**.

[25]See **3.9 Scale, 3.8 Fractals, 3.7 Chaos** and **14.13 Aggregates**.

[26]See **10.4 Computer Simulation**. Further reading Maddox (1992b).

[27]This is discussed further in **1.13 Phases, 5.8 Phase Changes** and **12.4 Phase Diagrams**, with examples in **4.4 Heisenberg May Have Been Here!, 11.6 Turbulence, 12.5 Mixtures, 15.8 Liquid Crystals, 15.5 Gels** and **14.7 Surfactants**. Further reading Wilson (1979), Scher *et al* (1991), Mehta and Barker (1991), Kadanoff (1991), de Gennes (1992).

[28]See **8.1 Atom Models**.

[29]Further reading Farmelo (1992), Ryder (1992), Cline (1994). The interaction of probes with materials is discussed further in **6.2 Probing Matter**, and in Chapter 7.

[30]The term 'fluid' means a material with 'ability to flow' and includes gases as well as liquids. Further reading on brownian motion Kaye (1993, Chapter 1). For colloids see **14.7 Surfactants** and **15.6 Colloids**.

[31]See **15.1 Soft Matter**.

[32]See **4.2 Wave States** and **6.4 Changing Matter**.

[33]Examples of limiting behaviour can be found in **3.2 Dynamics, 4.1 Ultraviolet Catastrophe, 5.3 Absolute Zero** and **9.5 Classifying Matter**. See also figure 3.2.

[34]Ruelle (1991, Chapter 25).

[35]Further reading Ruelle (1991, including Chapter 10).

[36]See, for example, **7.7 Quantum Dots**.

[37]Further reading Kaye (1993, Chapter 6).

[38]Similar situations occur with all materials: examples are provided in **2.4 Condition Critical** and **11.6 Turbulence**. See also **3.8 Fractals**. Further reading Mandelbrot (1983, p 17), Gleick (1988, p 99).

[39]Discussed in **6.1 Interactions**. Further reading Davies (1985, Chapter 4).

[40]Further reading Davies (1980, Chapter 2).

[41]Further reading Hawking (1988, p 163), Hawking and Penrose (1996).

[42]See **3.1 States, 11.5 Glasses** and the discussion on quasicrystals in **13.2 Crystals**. Further reading Elliott (1991, p 452).

[43]There is a link here to chaos: **3.7 Chaos**. Further reading Traub and Woznikowski (1994).

[44]Further reading Amato (1992d), McPherson (1989).

[45]Further reading Heilbronner and Dunitz (1993), Postle (1976, Chapter 10).

[46]Further reading Schattschneider (1994).

[47]See **9.14 Fullerenes**. Further reading Hammond and Kuck (1992, p ix), *Journal of Chemical Education* (1992), Boo (1992).

[48]The term 'molecular particle' is used here for either an isolated atom or a molecule, because some materials (**8.7 Noble Gases**) do not usually form molecules and remain in the atomic form in bulk matter.

[49]See **3.5 Flow, 3.8 Fractals** and **13.2 Crystals**. Further reading Stewart (1992).

[50]See **4.5 Playing Cards and Four-Strokes**. Further reading Postle (1976, Chapter 10).

[51]See **9.12 Handedness** and **5.4 Energy and Entropy**.

[52]Further reading Davies (1985, Chapters 4 and 7).

CHAPTER 3

PATTERNS IN MATTER

Before we can make effective use of our images or models of materials, we have to develop our own conceptual framework to accommodate those models. We have to clarify the relationships between them, and explore their similarities and differences. The models are interdependent, with various links between them, so it is not possible to choose a single, logical order of presentation that will satisfy us all. Rather, it is necessary to develop the concepts gradually, so that we can absorb them at our own pace, and eventually integrate them into our own conceptual model of matter.

3.1 States

The 'state' of a system is a formal description of the essential information about a system. It is the minimum information necessary to describe it fully. For example, it turns out[1]† that once we know the temperature, pressure and volume of a given 'amount'[2] of a pure substance we have enough information to specify its 'equation of state'. Also in common use is the term 'state of matter'. Its usage is less well defined than the 'state' of a system. We use 'state of matter' in referring to the properties of a distinct phase like ice or water. 'Change of state' is the conversion of one phase into another (such as the change of ice into water). 'Changes of state' are also known as 'phase changes'

† Endnotes for this chapter can be found on page 76.

or 'phase transitions'. The nature of a material and the identity of the particular state which is stable at a given temperature can be discussed in terms of an energy model. Energy is closely related to the group of properties, pressure, volume and temperature[3].

We can present this state information mathematically. However, to visualise it more clearly we can use the model of 'state space' or 'phase space', by extending our idea of 'real' (three-dimensional) space. State space can have more than three coordinates, or different coordinates chosen to suit the particular system. For physical systems it is often convenient to choose position and velocity coordinates, or position and momentum (momentum being velocity multiplied by mass)[4].

3.2 Dynamics

We also need to know how the state of a system evolves or changes with time ('temporal evolution' is the usual jargon) and we use 'dynamics' to model this process. As time passes, we can describe how a system evolves from a particular initial state by the way it moves through a path (or 'orbit') in state space. We can either indicate the behaviour continuously as a function of time (using a 'flow diagram') or obtain a 'snapshot' of the situation at a particular time.

Mathematically, we can solve equations to describe the state of the system as a function of time. Occasionally, in the right circumstances (usually when we have idealised a real life situation into a highly simplified model) we can obtain mathematical solutions in 'closed' or 'integrable' forms (mathematical equations that we can integrate)[5]. Simple equations then predict behaviour at any given time in the future. Examples are

- a frictionless pendulum (or a pendulum in a clock where the driving mechanism offsets frictional forces)

- the 'classical' model[6] for motion of the planets around the Sun.

No exact solutions[7] are possible for most real systems, such as weather patterns in the atmosphere, no matter how much information we have. (The 'butterfly effect' suggests in the extreme case that the fluttering of a butterfly can eventually change the state of the atmosphere on the other side of the Earth.) The long-term evolution of such systems is unpredictable because of the extreme sensitivity to initial conditions, as we see below[8]. There are, however, numerous situations where interaction of various events causes a steady state situation, with the macroscopic or large-scale conditions predictable despite the random nature of the numerous minor events causing them. A feedback mechanism, with information

on the state of the system modifying subsequent processes, has
to be involved[9].

With the state space model we can visualise the behaviour of a
system in geometric form. In some systems, orbits in state space
are attracted to particular regions ('attractors'), and the
identification of these regions greatly facilitates our understanding
of the systems. Attractors (which can be points, lines, planes or
multidimensional surfaces) are where systems tend to rest. Tom
Mullin[10] uses the analogy of a ball bearing rolling around inside a
tyre inner tube. In any two-dimensional cross-section of the tube,
the ball bearing always comes to rest at a particular point at the
bottom of the tube: the particular attractor in this system is a
point. If we place the inner tube flat on a turntable rotating at
constant rate, the ball bearing will move around the tube but stay
somewhere on a circle. This circle is now the attractor.

We can represent the motion of a pendulum in state space, for

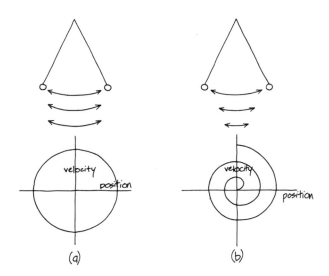

Figure 3.1 Representations of the behaviour of (a) a frictionless
pendulum (or a pendulum driven by a clock mechanism) and (b) a
pendulum coming to rest. The circle in (a) and the central point in (b)
are the state attractors.

example, by plotting its position against its velocity (speed in a certain direction). When a real pendulum is part of a clock, which we are trying to 'start' by swinging the pendulum, small initial displacements of the pendulum do not result in the clock 'starting'. Because of the mechanical friction the pendulum eventually comes to rest at its lowest position. This point is the state space attractor for the system (figure 3.1(b)). The orbit of the pendulum oscillating with ever-decreasing amplitude and eventually reaching the attractor point can also be represented in state space. With a larger initial displacement, the clock begins to 'tick' and the pendulum undergoes a stable oscillation. The region of state space representing this stable oscillation is the second attractor in this system. It takes the form of a cyclic orbit in state space, and represents the limiting behaviour that the system develops (figure 3.1(a))[11].

The 'relaxation time' for the characterisation of a dynamic process (introduced previously[12]) is related to the 'residence time'. This is the characteristic time spent by a representative particle in a system or in a portion of a cycle. Each gaseous component of the atmosphere (such as carbon dioxide in the cycle of carbon through plants, perhaps coal or animals, the atmosphere, the oceans and carbonate rocks) has a residence time in the atmosphere. The residence time tells us the extent of interaction of this gas with the other chemical and physical and biological parts of the Earth[13].

3.3 Newton's Model

More than three hundred years ago Isaac Newton established a model for the behaviour of objects and forces in time and space, summarised in 'Newton's laws of motion'. This simplification or model of physical reality is still widely used and (with some limitations) still relevant. An object in this model continues in its state of rest or at constant velocity unless acted upon by an external force. The rate of change of momentum of the object is proportional to the force causing it and takes place in the direction of that force.

Here it is necessary to note that velocity is a 'vector' quantity, meaning speed in a particular direction. Momentum is the product of mass and velocity, so force and momentum are also vectors. Although mass is not a vector quantity, the weight of an object (the gravitational attraction between the object's mass and the Earth's mass) **is** a vector, directed towards the centre of the Earth. Vectors are mathematical models facilitating the combination or separation of effects in any number of dimensions[14].

This approach is 'determinism': the prediction of future positions of objects from past observations, a philosophy that was unquestioned until recently[15]. Newton provided not only a conceptual model (a way of thinking about the way things work) but also a mathematical model, permitting the prediction of quantitative results very simply. Gravitational attraction is proportional to the product of the masses and inversely proportional to the square of the distance between them. This model of mechanics, described as 'newtonian' or 'classical', is still the basis for extremely complex and lengthy calculations carried out in computer simulation[16].

Although it may not be immediately obvious because of our experience and conditioning, the two concepts of 'mass' used in these models differ:

- One is an 'inertial' mass, a 'kinematic' quantity having to do with the motion of an object (associated with the force necessary to accelerate it).

- The other is a 'gravitational mass', in modern terms a sort of 'charge' that determines how a massive object reacts to a gravitational force. (The word 'massive' here means 'having mass' rather than 'very heavy'.) This is rather like the way an electric charge determines the reaction to an electromagnetic force.

Newton originally checked the equivalence of the two mass concepts to one part in a thousand. It was subsequently verified to a billionfold higher precision by others, including a Moon-based experiment installed by the Apollo astronauts[17].

3.4 Fields

A model valuable in some areas of science assumes that a body (or charge) sets up a field of influence with which any other body (or charge) may interact. This is an idea introduced by James Clerk Maxwell and Michael Faraday. (In the first chapter, I asked you to carry out the thought experiment of 'creating' a particle from 'empty' space by distorting and concentrating it. If you now consider a space formed into tiny 'particles' of energy moving at light speed, 'photons', you have created an electromagnetic field.) When people had difficulty in visualising a field existing in a vacuum, they borrowed the Greek 'ether' to provide a notional medium in which the field existed. In retrospect, we can see that the ether idea held up the development of the wave model[18].

To use the field model quantitatively, we define a 'field strength', with its magnitude at any point determining the extent of interaction with any body (or charge). Some of the most common fields are:

- gravitational field near any object possessing mass
- electric field near an electric charge
- magnetic field surrounding a pair of magnetic poles ('dipole').

We can describe almost any physical property (such as temperature) with a field. Maxwell in the middle of the nineteenth century made great advances with a mathematical form of the electromagnetic wave model based on the idea of interacting electric and magnetic fields. This was the forerunner of modern field models[19].

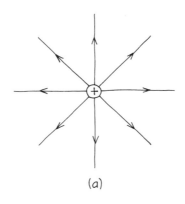

(a)

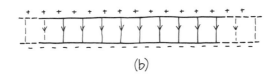

(b)

Figure 3.2 Field lines associated with (a) point positive electric charges and (b) parallel charged plates carrying opposite charges.

The value of the field model is that we can analyse an interaction by dividing it into two parts:

- a 'condition' or effect produced by the source of the field at a particular place, and

- the property of a second mass, charge or magnetic pole that 'feels' or experiences this condition.

The mass of the Earth produces a condition at the surface of the Earth that we describe by the 'gravitational acceleration' or 'gravity'. Another object near the surface of the Earth experiences this condition and interacts with it, resulting in the object being accelerated in the direction of the centre of the Earth. One part of this model says that something produces a field, and the other part says that the field acts on something else. This separation allows us to view the two parts independently, and

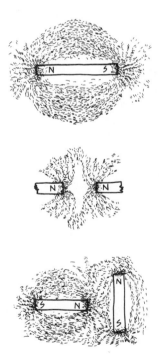

Figure 3.3 The patterns in iron filings caused by magnets suggested the use of field lines to model magnetic interactions.

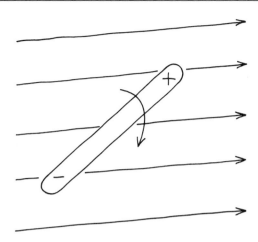

Figure 3.4 Rotation of an electric dipole in an electric field.

so simplify the calculations as well as providing a useful conceptual model. If, for example, several electric charges are present, we may calculate or measure the total electric field produced by them at a certain point. Subsequently, if we place another known electric charge at that point, we can calculate from the field strength the nature and extent of the resulting interaction.

In the field model we use field 'lines' to indicate the strength and direction of the field. For a compact source of charge, for example (figure 3.2(a)), the lines radiate outwards from the charge. They are closer together near the charge, showing that the field intensity is greater there. (By convention, the 'direction' of the field lines is outwards from a positive charge, the direction that a small, positive test charge would take if introduced into the field.) Between parallel charged plates (figure 3.2(b)) the field lines are parallel to each other, and (away from the edges of the plates) they are evenly spaced. We can readily illustrate magnetic field lines by placing iron filings on a card just above a magnet, and tapping the card to enable the particles to align themselves in the magnetic field (figure 3.3). An electric dipole in an electric field experiences a rotating force or torque (figure 3.4).

However, it is important to remember that field lines are part of this model devised to assist us to visualise the 'invisible' phenomenon existing around the charges or between the magnetic poles. We choose for ourselves how close to draw the lines to denote a particular field strength, just as on a map we can choose how far apart we draw the contour lines to represent changing heights. In figure 3.5, on the western side the island slopes more steeply, so the contour lines are closer together there. Each particular configuration of electrical charges or magnetic poles or gravitational masses has its own corresponding field. Where possible we tend to deal with limiting behaviour configurations with a high degree of symmetry (like figure 3.2(a)) because they are easier to describe and to visualise. However, real, complex systems (like gravitational effects in the Earth–Moon system, figure 3.6) can be depicted easily.

Electric and gravitational fields are 'vector fields', because direction is important. Other fields such as temperature are 'scalar fields': there is only magnitude associated with each point.

In the field model, a field can transfer energy to a body. For example, a stone falling in the Earth's gravitational field gains kinetic energy and to balance this we refer to a decrease in the stone's potential energy in the field. Between 1846 and 1851 William

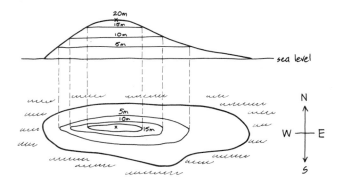

Figure 3.5 Plan and elevation of an island, with contour lines at 5 metre intervals. The closer the lines on the plan, the steeper the slope.

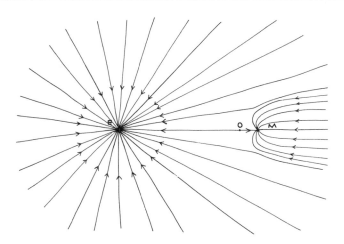

Figure 3.6 Gravitational field lines for the Earth (E)–Moon (M) system. Note point O, where the resultant gravitational field is zero.

Thompson, Lord Kelvin moved from the view that **forces** were most significant to one that made **energy** central in science, and this emphasis has endured. Energy may seem real enough to us, because it is a word and concept in widespread everyday use. Although energy is a valuable model it is not an obvious or simple one[20].

We also use the term 'field' in the context of quantum electrodynamics to describe the zero-point radiation which is always in existence in conjunction with matter[21]. We can extend the field model still further. We can think of the various kinds of fundamental field (gravitational, electromagnetic, and so on) as being permanently and simultaneously in existence throughout space. We can also consider that the objects we choose to call 'particles' are themselves the results of interactions between these fields[22].

3.5 Flow

Another model useful in some situations is the 'field of flow'. If there is a difference between the potential at two points (that is, if there is a potential gradient), flow can occur. For example, we can consider the flow of a liquid through a channel under a pressure gradient. We draw field lines or 'streamlines' to represent the path that small test particles would take if introduced into the flow. Each point has associated with it a vector quantity (a quantity where direction is important). In the case of flowing water it is velocity. The rate of flow with time is the 'flux', and electric flux under an electrical potential gradient is simply electric current. Similarly, materials flow or 'diffuse' under a concentration gradient, and heat flows under a temperature gradient. (This is the most fundamental definition of temperature. Temperature is that property that defines the **direction** of energy flow as heat from higher to lower temperature[23].)

In each of these flows we often observe experimentally that the flux (rate of flow) or flux density (flow per unit cross-sectional area in unit time) is proportional to the potential gradient. The proportionality factor or 'transport coefficient' is a characteristic of the medium in which the flow is occurring. In electrical flow this transport coefficient is the 'electrical conductivity', in heat flow it is 'thermal conductivity' and in diffusion of one material through another it is the 'diffusion coefficient'. In all these examples there is a flow of energy or matter through a stationary medium. Fluid flow, whether the fluid is a gas, liquid or other material, differs in that the whole medium is flowing.

3.6 Light

Our inadequate ability to observe tends to condition our view of the world, and this is true even of light itself. Prisms (or water droplets in a rainbow) split what we normally see as 'white light' into 'colours' of the visible spectrum. But in reality these 'colours' extend in one direction beyond red to infrared and microwave and radio waves, and in the other direction beyond violet to the ultraviolet, x-rays and gamma rays. The wave model regards light as a continuous array of 'wavelengths' or 'frequencies'. We can also split light in a different way by using a polariser[24].

Isaac Newton considered that light was made up of 'corpuscles'

('photon' is the modern term). Even though we may not agree
with the way he reached his decision, this model is consistent with
what we now know about the behaviour of light. Physicists have
instruments (photomultipliers) that enable one photon to
generate a significant and measurable electric current, and we
know that one photon cannot be detected at 'half strength' by two
photomultipliers. Richard Feynman[25] has pointed out that if our
own eyes were about ten times more sensitive than they are, we
would see a dim light as a series of flashes as each individual
photon activated a nerve cell. If this had been the case and we
could have 'seen' single photons, we might have been satisfied
with the particle model and not gone on to develop a wave model
to fit some of the other properties of light. Nevertheless, as
demonstrated by Thomas Young early in the nineteenth century,
there are situations where it is convenient to treat light as a wave.
This is similar to regarding matter either as continuous, or as an
assembly of atoms, or as a wave, depending on the particular
circumstances[26].

We should accept our imperfect ability to visualise reality and
consciously make full use of whatever models are most helpful to
us in improving our understanding of it.

3.7 Chaos

One of the useful 'model' classifications of systems of materials is
based on the degree of order they exhibit. At one limit, materials
are ordered; at the other they are random or disordered. In
intermediate situations we must ask the extent of order.

- An ideal crystalline solid, with a high degree of symmetry, is
  an ordered material.

- A gas is a good example of a random or disordered
  arrangement, being isotropic (appearing the same from all
  directions) and without structure.

- Perfect crystals demonstrate long-range order, while in liquids there is some short-range order imposed by the packing arrangements.

Without specifically analysing the degree of order, or structure, or symmetry, we can assess the natures of states of matter on statistical evidence, a discipline known as 'statistical mechanics' or 'statistical thermodynamics'. You have an ability to distinguish between the styles of artists like van Gogh and Cezanne or between the music of composers like Bach and Beethoven, intuitively using probabilities to combine particular characteristics. This ability models[27] the way we can apply statistical methods to collections of atoms or molecules to determine bulk properties. If we apply restrictions to the large-scale, macroscopic state of a system, we also largely control the small-scale, probabilistic structure. The macroscopic condition often specified is the **temperature**.

Some systems are not only random but also 'chaotic', characterised not just by uncertainty, which is the case with random systems, but also by an exponential amplification of uncertainty as time passes and the system evolves[28]. The exponential rather than linear growth of uncertainty with time in a chaotic system soon makes its state unpredictable. University of Maryland applied mathematician James Yorke in the mid-1970s chose the word 'chaos' to have this special meaning. Long-term unpredictability as a result of a small initial uncertainty was considered in detail by Henri Poincaré early in the twentieth century, but the model took time to become accepted in science. It does appear that the introduction of the word 'chaos' promoted popular acceptance of the model[29]. The test for the existence of chaos is whether the collection of more information can in principle achieve predictability in the system. We can show that even some systems with only a few elements are unpredictable. Nevertheless, not everything about a chaotic system is necessarily unpredictable. Those systems that return repeatedly to a previous state (those showing 'eternal return') provide the opportunity for a kind of predictability: we can explore other forms of order or patterns in chaos. Order can arise spontaneously as the result of chaotic natural events. Living systems are particularly efficient at maintaining properties which are roughly constant and predictable (the body temperature of mammals, for example) while responding to unpredictable events around them. Another dramatic example is the Red Spot of Jupiter, which has similarities to the occasional cases of 'blocking' by stationary anticyclones in our weather patterns. These are self-organising systems in the midst of chaos[30].

The trajectory of a cricket ball after its collision with the bat is predictable. Gravitational interactions (and to some extent interactions with molecules in the air) modify its path in a predictable way. But this predictability does not exist in a series of collisions between convex objects of similar size, for example between billiard balls. Because the surfaces are curved, small differences at the points of impact are amplified exponentially with each collision, and the final state becomes unpredictable. The same argument applies to the behaviour of the atoms or molecules in a gas, even in a model 'gas' composed of 'hard spheres'[31]. It is a waste of time even in this system composed of classical particles to define initial positions or initial velocities more and more precisely in an attempt to predict the final outcome. In such systems there is a sensitive dependence on initial conditions. Scientists use terms such as 'weak stability' and 'strong mixing' to describe them. There are some cases where this sensitivity is a matter of 'common sense' (such as sensitivity to the particular condition in trying to balance a pin on its point). There are other systems (such as any series of billiard ball impacts) where the sensitivity to arbitrary initial conditions is much less obvious. The result may be intuitively unexpected or counterintuitive until you have thought it through carefully[32].

There are ultimate limits on our ability to define the properties of systems at an atomic level[33], and chaotic processes amplify this uncertainty. The uncertainty exists even at low temperatures: it is not thermal 'noise' that we can remove by lowering the temperature.

The model of 'state space' is useful in visualising the distinction between order, randomness and chaos. When we focus our attention on a small region of state space, in a nonchaotic system those points that start off close together remain close as the system evolves. If chaotic behaviour occurs, nearby points diverge rapidly. This divergence cannot persist indefinitely (because of the finite extent of the system) so we can visualise the system as being folded back on itself, bringing close together points that were once well separated. Chaos causes the 'mixing' of state space; the model of kneading dough has been used[34] to illustrate in the three dimensions of real space what is happening in state space. A small amount of colouring matter introduced into the dough stretches and folds until it becomes rapidly distributed at a molecular level throughout the mixture[35].

'Fractal clusters' (below) of soot formed by the aggregation of small particles have their own type of order or pattern, even though they have as their origin interactions between unrelated events in apparently random systems. In flowing fluids (liquid or gas), 'laminar flow' is ordered, while 'turbulent flow' is chaotic. (You can observe the transition between laminar or streamline flow and turbulent or chaotic flow by gradually increasing the flow from a water tap.) Often we wish to ensure that there is laminar flow in pipes to minimise energy requirements, but the mixing of fluids is a situation where turbulence (chaos) is beneficial[36].

Chaotic behaviour does not require the involvement of a large number of particles. A system as simple as a one-dimensional pendulum subjected to a periodic perturbation exhibits chaotic behaviour. The model of a 'damped and driven oscillator in an asymmetric double-well potential'[37] fits various physical and chemical systems. Although we can describe the gravitational interaction of two masses such as two planets exactly, the presence of a third smaller planet makes the behaviour of the system ultimately unpredictable[38].

We find that apparently predictable systems are chaotic over a long period. A recent computer simulation of a one hundred million year period in the life of the solar system (a period far shorter than its past five thousand million year history) showed that the orbits of the planets are chaotic, with unpredictable wandering. However, there is also evidence of constraint that prevents catastrophic behaviour, unlike the behaviour of some of the smaller bodies (such as asteroids) in the solar system[39]. The concept of 'constrained chaos', with unpredictability between certain limits, is likely to prove important in a variety of situations. An example is asteroid Helga, strongly influenced by the planet Jupiter, with an orbit that is clearly chaotic (its position becoming unpredictable after seven thousand years) but which in the model calculations was found to remain in orbit for at least seven million years.

The sensitivity of chaotic systems to small perturbations can be utilised in practical systems, for example by modifying chaotic trajectories of spacecraft with minimal fuel use. (Here **large** changes result from small controlling actions in such chaotic situations. In contrast, a long ruler balanced upright on one's upturned hand can be stabilised in an unstable steady state by making small corrections as long as the ruler does not deviate excessively from the vertical. This is nonchaotic because **small** correcting effects can be achieved with small actions[40].) Systems based on chaos can be used to control and stabilise mechanical, electrical and biological systems[41].

It is comforting that although nature is not simple and predictable, neither is it completely random and unpredictable. It is between these extremes that natural phenomena tend to exist, with repeating patterns which are independent of scale: scale-free patterns or fractals, generated by self-organised criticality.

3.8 Fractals

The term 'fractal' was created by Benoit Mandelbrot in 1975 from the Latin, implying 'irregular fragment'. Although is of very recent origin, it has already been used in several ways, ranging from a mathematical definition to very irregular natural phenomena. All uses suggest similar structures nested one within another like Russian dolls: systems with properties that repeat at innumerable scales. 'Fractal' is another word whose introduction has popularised a model. Fractals are characterised by having a similar appearance regardless of the scale of enlargement of the pattern or design, despite the absence of 'order', and are associated with chaotic behaviour. Fractal behaviour and structure are now being recognised in more and more situations. A typical natural fractal is a particle of soot, with an intricate structure visible under magnification, formed by 'diffusion-limited aggregation'[42] that is tenuous rather than compact. ('Tenuous' is a word meaning 'insubstantial' or 'flimsy': the opposite of 'close-packed'.) Less obvious but just as impressive is the efficient fractal system of blood vessels in our bodies, which show a continuous branching structure (which for convenience we classify in an arbitrary way as 'arteries', 'veins', 'capillaries'). The structure of a tree is similar: trunk, branches, twigs are only approximate descriptions of a fractal structure. We now have 'fractal chemistry': the synthesis of tree-like, dendritic polymer molecules[43].

'Brownian motion'[44] also illustrates the properties of fractals, as it is a 'random walk' where we can use the model of a drunkard performing a series of random motions in all directions. (We can apply this model also in diffusion and in diffusion-limited aggregation.) We see the same kinds of movement microscopically no matter how small a scale we study them on, so these are further examples of natural fractals. If we plot the path of one particle subject to brownian motion, the trace (nominally a 'one-dimensional' curve) eventually will fill the whole page, and be of dimension two[45].

The fractal model is just as relevant to molecular branching as it is to the structures of the airways in our lungs, or trees, or coral reefs. It is another 'way of thinking about the dimensional hierarchy of the Universe'[46].

3.9 Scale

The self-similarity of fractals is formally described in terms of 'scale invariance'. The same relationships apply regardless of the time scale of the observation, or of the units being used in distance scales. For example, the general shape of a coastline is independent of the scale, so that when we compare an aerial photograph with an enlarged section of the same photograph, we cannot distinguish the two by general appearance[47].

It is becoming clear that often the outcome of random events (such as radioactive decay) is based on the terms of a geometric series (such as x, x^2, x^3, ...) rather than on an arithmetic series (such as x, $2x$, $3x$, ...) and that this leads to scale-invariant behaviour. One of the frequent results is that when we measure things on a linear, arithmetic scale using our base-ten notation the frequency with which a digit occurs in the initial position of any number decreases in a regular way as the digits increase from 1 to 9. In other words, there are far more numbers of things starting with 1 than starting with 9. This has the intuitively unexpected result that the **first digits** of many large, naturally occurring sets of numbers when collected in decimal form (that is, written as ordinary numbers, 0, 1, 2, etc) are not uniformly distributed. Instead, they follow a logarithmic probability distribution, with the likelihood of one as the first digit being greater than that of the first digit being two, and so on. More than 30% begin with the digit 1 while less than 5% begin with 9. This observation is 'Benford's first-digit law'[48].

3.10 Dirty Systems

Traditionally, scientists have preferred to study 'clean' or simplified systems. Physicists go to extreme lengths with high-vacuum systems to obtain clean surfaces. Chemists have tended to study dilute solutions, which are perceived as being 'simpler' than concentrated solutions. Techniques such as x-ray crystallography reflect the average, ideal bulk structure of solids rather than the dislocations and imperfections. The liquid state has not been studied as extensively as gases or solids because it does not have the convenient randomness of the former or crystalline order of the latter. Awareness of the importance of confronting the problems posed by 'dirty' systems, combined with new experimental techniques and modelling methods, enable us to make progress in these areas. Sometimes there are properties we can observe only in the presence of impurities, such as 'semiconductor' systems and some types of superconductor[49]. 'Spin glasses' are the ultimate in 'dirty' systems, as we see in a later chapter[50].

Endnotes

[1]Discussed in **5.2 Thermodynamic Model**, **12.4 Phase Diagrams** and **12.1 Purity**.

[2]The word 'amount' in chemistry has a particular meaning: see **9.1 Amount**.

[3]See **1.13 Phases** and **5.8 Phase Changes**.

[4]See **2.6 Limits and Dimensions**. 'Velocity' means the same as speed, but 'speed in a specified direction', and momentum is given by the product of velocity and mass: see **3.3 Newton's Model**.

[5]The mathematical process of integration allows information on how a property is **changing** over time to be converted into one that specifies the value of the property at any particular time.

[6]See **3.3 Newton's Model**.

[7]No 'closed' or 'integrable' solutions can be found.

[8]See **3.7 Chaos**. Further reading Kaye (1993, Chapter 1).

[9]Examples are given in **5.7 Ordering**.

[10]Mullin (1993).

[11]See **2.6 Limits and Dimensions**. Further reading Crutchfield *et al* (1986), Mullin (1993, p 2), Ditto and Pecora (1993).

[12]See **2.2 Perspective**. The relaxation time is important in viscoelastic properties, see **11.4 Viscous Flow** and **11.5 Glasses**.

[13]See **15.12 Life**. Further reading Lovelock (1988, 1991). Time dependence is discussed further in **5.9 Dynamic Models**.

[14]**2.6 Limits and Dimensions**. Further reading Davies (1980, Chapter 3).

[15]See **4.4 Heisenberg May Have Been Here!** and **3.7 Chaos**. Further reading Kaye (1993, Chapter 1).

[16]See **10.4 Computer Simulation**.

[17]See also **3.4 Fields** and **7.8 Colourful Quarks**. Further reading Goldman *et al* (1988).

[18]Further reading Arnau *et al* (1995).

[19]Discussed in **4.2 Wave States** and **7.8 Colourful Quarks**. Further reading Davies (1985, Chapter 4).

[20]Discussed in **5.1 Energy Well Model**, and also **4.6 Quantum Leap**. Models based on virtual particles, **7.9 Virtual Reality**, can also be used in the discussion of the influence-at-a-distance of particles, but for everyday situations field models are more useful. Further reading Davies (1985, Chapter 4).

[21]See **4.4 Heisenberg May Have Been Here!**.

[22]This concept is explored further in Chapter 7. Further reading Postle (1976, Chapter 5).

[23]Discussed in **5.2 Thermodynamic Model**.

[24]See **1.3 Galloping Horses** and **4.2 Wave States**.

[25]Feynman (1985, Chapter 1).

[26]This is discussed further in **4.1 Ultraviolet Catastrophe** and **4.2 Wave States**. Further reading Jones (1991), Horgan (1992).

[27]Ruelle (1991, Chapter 19).

[28]Further reading Ruelle (1991), Berry *et al* (1987). For exponential functions, see **2.2 Perspective**.

[29]Further reading Ruelle (1991, Chapter 11, 1994), Gleick (1988), Mullin (1993).

[30]Further reading Gleick (1988, p 53).

[31]See **10.6 Hard Sphere Model**.

[32]Further reading Ruelle (1991, Chapter 7), Rae (1986, Chapter 9), Mullin (1993, p xi).

[33]The Heisenberg uncertainty principle, **4.4 Heisenberg May Have Been Here!**, **4.1 Ultraviolet Catastrophe** and **8.4 Quantum Chaos**.

[34]Further reading Crutchfield *et al* (1986, p 43).

[35]The state space attractor for chaotic behaviour is not a point or a limit cycle but a 'fractal' object (a 'strange attractor'): an object which reveals ever more detail as it is increasingly magnified and which is similar on all scales. See also **3.8 Fractals** and **14.13 Aggregates**. Further reading Mandelbrot (1983), Kadanoff (1986), Stewart (1992), Ruelle (1991), Mullin (1993).

[36]Discussed in **11.6 Turbulence** and **12.5 Mixtures**. Further reading Ottino (1989), Ottino *et al* (1992), Gleick (1988).

[37]Further reading Tagg *et al* (1994).

[38]Further reading Mullin (1993, p 2). The simple harmonic motion of a pendulum is an ideal or model limiting behaviour at very small displacements that can be described by linear equations, but the behaviour of a real (nonlinear) pendulum is much more complex.

[39]Further reading Kerr (1992).

[40]Further reading Shinbrot *et al* (1993).

[41]Further reading Ditto and Pecora (1993), Ruelle (1994).

[42]Discussed in **14.13 Aggregates** and **14.10 Aerogels**.

[43]See **9.15 Macromolecules**.

[44]See **2.5 Smaller and Smaller Probes**.

[45]This is an expansion of the concept discussed in **2.6 Limits and Dimensions**, and

prompts us to think about our perception of 'dimension'.

[46]Tomalia (1995). Further reading Donn (1990), Mandelbrot (1983), Kadanoff (1986), Stewart (1992), de Gennes (1992), Gleick (1988, p 81), Kaye (1993), Dewdney (1987, 1989), Bak and Paczuski (1993), Weber (1992, p 22).

[47]See **2.2 Perspective**. Further reading Cardy (1993), Finnis (1993).

[48]Further reading Stewart (1993), Buck *et al* (1993), Raimi (1969). A dramatic demonstration is the radioactive decay of the nucleus (**7.10 Isotopes**) by the emission of alpha particles, where the probability of escape depends exponentially on a function of energy differences, which are likely to be randomly distributed depending on the type of nucleus. It was found that the distribution of the half-lives of nearly 500 different types of alpha decay, spanning 28 decade orders of magnitude of time between a microsecond and 10^{15} years, follows a logarithmic distribution.

[49]See **6.2 Probing Matter** and **13.7 Superconductors**.

[50]Discussed in **13.9 Magnetic Matter**. Further reading Stein (1989).

CHAPTER 4

UNUSUAL MATTER

Up to this point, the models and concepts described have not been all that strange or unusual. Most are closely linked to our sense of physical, everyday reality. Even chaos as a scientific concept does not seem all that unreasonable. But now I come to ideas that require you to set aside some of your preconceptions about the essential nature of things. You will have to be able to accept 'quantisation' and the 'wave' nature of matter, the unusual symmetry properties of particles and the fundamental uncertainty about everything.

4.1 Ultraviolet Catastrophe

According to the 'classical' model of energy transfer that we accepted until early in the twentieth century, hot objects should radiate energy at all wavelengths (or frequencies). In particular, there should be large amounts of energy at high frequencies: the 'ultraviolet catastrophe'. Because a classical object should radiate energy at all frequencies and because the number of frequencies is in principle unlimited, the rate of energy radiated would be infinitely great, which is clearly unreasonable[1]†.

At the end of the nineteenth century, we 'recognised' 'particles' of matter (atoms) and 'particles' of electricity (electrons). By 1905 Max Planck and Albert Einstein had extended the idea to energy,

† Endnotes for this chapter can be found on page 99.

proposing that energy could be radiated only in packets of fixed amounts of energy, 'quanta'. (*Quantum* is the Latin word for 'amount'). These fixed amounts became higher as the frequency became higher, ultraviolet radiation having a higher quantum energy than visible or infrared radiation. At a sufficiently high frequency one quantum would require more than the available energy, so providing a natural limit to the rate of radiation and avoiding the ultraviolet catastrophe. (Gilbert N Lewis subsequently named the 'particle' of electromagnetic radiation the 'photon', twenty years after the original suggestion that light was quantised.)

We notice that objects in everyday life ('classical objects' rather than 'quantum objects') are able to have any value of energy: there appears to be a continuous range of possible energies. Quantum objects or quantum particles, however, can have only certain discrete energy states and the energy is 'quantised'. (From now on I shall use the term 'quantum particle' rather than 'quantum object'. Although objects of all sizes have quantum characteristics, we usually see departures from classical properties only in very small objects, and 'particle' reminds us of this.)

For the wave model, we can think of the pattern of wavy lines sometimes used as a decoration around the edge of a plate. If the pattern is to be continuous, if the lines are to join up properly where they meet, there can be only a whole number of waves around the plate and not intermediate fractional waves. Another analogy is the way we get money from a bank automatic teller machine (ATM). From the perspective of a mechanical cash dispenser, money is quantised in one or two denominations of notes rather than the lower-valued single-coin quantisation of the rest of our transactions[2].

In the classical electromagnetic model, an atom with a positive nucleus and orbiting negative electron[3] is not feasible. But as quantum particles, electrons in atoms can lose energy only by 'jumping' from a higher level to a lower level. As they do, they give off the difference in energy as electromagnetic radiation, and from the lowest level or ground state no further loss of energy is possible. The popular terms 'quantum leap' and 'quantum jump' have the unfortunate connotation of a **large** energy jump. A quantum of energy does imply a 'jump', but it is a very small energy jump by our usual standards.

Quantum mechanics or wave mechanics is a mathematical model devised by Erwin Schrödinger in the mid-1920s, with contributions from Satyendra Nath Bose (see below), Louis de Broglie, Albert Einstein and others. It is applicable to situations where quantum particles are subject to a force. The most important situation of this kind in the atomic model is that of the electron under the influence of the positive nucleus. Quantum models predict that an atom must occupy one of a particular set of energy levels or states, the amount of energy depending on its electronic state and its state of motion. An atom can acquire energy by collision with another particle or by absorbing electromagnetic radiation, but only in set amounts.

Materials at molecular, atomic and subatomic levels have properties that do not conform to our understanding of everyday classical behaviour. Quantum models are able to describe these properties as of wave functions (mathematical expressions with values varying in both space and time). The 'Schrödinger equation' describes quantum particles like electrons by means of wave functions.

Some experiments show that these quantum particles behave as we would expect 'ordinary' particles to behave. Others show that they can also behave as waves, with interference effects arising from the adding or cancelling of troughs and crests. Practical experiments are now being devised (like the 'mind' experiments proposed by Werner Heisenberg, Niels Bohr and Albert Einstein in the late 1920s) to investigate the 'wave–particle duality' not only of photons and electrons, but also of atoms. The wave model and the particle model are just two aspects (each of which on its own seems natural and reasonable) of a more complex totality which may be beyond our ability to visualise. The total picture has to extend all the way from subatomic matter to white dwarf stars and black holes[4].

We have to accept 'counterintuitive' quantum properties that do not necessarily conform to everyday observation or 'common sense'. For example, a photon is in an ambiguous state, either particle-like or wave-like, until we make an **irreversible** observation or specify an experimental arrangement. This is so even if we can make the experimental choice **after** an individual photon has interacted with the equipment. Photons are neither waves nor particles (these are just our convenient models for the description of aspects of their behaviour) but are undefined until we measure them. We cannot know whether quantum particles exist (whether they are 'real') until we make some kind of measurement. When photons interact with a 'diffraction grating'[5] they 'are' waves. When they react with silver ions in a photographic film they 'are' particles. And without such measurements it is meaningless to discuss their significance at all. Also surprisingly, two quantum particles separated by macroscopic distances appear to be 'entangled', such that correlations in behaviour occur, with a measurement on one affecting the result of a measurement on the other[6].

It is worth pausing to think about the way in which wave–particle duality was 'accepted' by scientists and eventually by the broader community. Planck developed quantisation of light as a mathematical model, and for twenty years he and most people still probably thought that light was 'really' a wave. One of the few people to intuitively accept that he could describe light completely in particle terms without even mentioning its wave-like properties was Satyendra Nath Bose of the University of Dacca. He treated photons as particles counted in a different way—obeying a different kind of group behaviour (statistics)[7]. When we say light behaves like particles but is really a wave, or that light is made up of particles but acts like a wave, we are introducing a conceptual barrier. Rather, I would like you to abandon the task of trying to say what light **is**, and instead say that we can model it both as a wave and as a particle.

The conceptual struggle that the scientific community had with wave–particle duality highlights our inability to accept more than one model at a time. Its eventual acceptance is an excellent example of the way we should approach all phenomena.

In an alternative model, the quantum mechanical 'particle' or 'wave packet' has been likened to a tortoise. The likelihood of the particle's occurrence rises from the leading edge to a maximum in the middle of the of the tortoise's shell, then slopes down again to the trailing edge. However, when we detect the particle in some way at a particular point, the rest of the wave packet (tortoise) is no longer relevant. This uncertainty in the position of the quantum particle leads to the idea of 'tunnelling'. If a quantum particle is bounced off a 'wall' or energy barrier, its trajectory is not definite as it would be for a classical object. There is a small but finite chance of it appearing on the other side of the wall. It has tunnelled through the energy barrier, just as Lewis Carroll's character, Alice, stepped through the looking glass barrier[8].

Most systems act in ways that tend towards those predicted by classical or newtonian models as their sizes become much larger than the size of atoms[9]. The 'correspondence principle' asserts that the upper limit of quantum behaviour is classical behaviour.

For consistency, it is helpful to visualise classical systems as still having the quantised energy levels. We must note, however, that these levels are so close together that ordinary atomic low-energy collisions result in energy transitions between levels. Therefore there appears to be a smooth distribution or continuum of energies. There is nothing unusual in this: it is analogous to the physical appearance of matter. A piece of steel seems to be continuous at the engineering or human scale, but fine structure appears under optical microscopy and at a higher magnification discrete individual atoms are detectable by scanning probe methods[10]. Once we have adopted this viewpoint it is easier to deal with those life-size systems that have properties not explicable in classical terms. For example, liquid helium has superfluid properties because even in macroscopic samples all its atoms are in the same quantum state rather than occupying a wide range of very close energy levels[11].

Our discussion on quantisation so far has been about energy quantisation. What about particles? The energy of a rotating particle is related classically to its 'angular momentum' (inertia associated with rotation), and from this we deduce that the magnitude of the angular momentum is also quantised. Further, this means that the orientation of a rotating body is quantised. 'Space quantisation' is the quantum mechanical result that a rotating body may not take up an arbitrary orientation in space. (This is referred to some specified axis such as the direction of an applied electric or magnetic field.) An experiment designed by Otto Stern and Walther Gerlach, in which a beam of silver atoms passed through a nonuniform magnetic field, confirmed this for quantum particles. Space quantisation is not detectable in macroscopic objects where the classical model is adequate[12].

As a model, classical mechanics is at first sight much more appealing than quantum mechanics because we can more readily apply our intuitive understanding of the ordinary world around us. Consequently, the boundary between classical and quantum behaviour has been explored from the point of view of extending classical methods to the limit, aided by the development of high-speed computers. 'Semiclassical mechanics' provides the link between classical and quantum mechanics. This involves the imposition of quantisation on classical quantities, and for chaotic systems the replacement of the intractable chaotic mathematical functions with nonchaotic approximations. Of particular interest recently has been the nature of the relationship between classical, quantum mechanical and chaotic systems and the correlations between quantum particles[13].

The uncertain nature of strong mixing in chaotic processes requires us to make a distinction between reversible and irreversible behaviour for both classical and quantum processes. Irreversible changes are normal and universal, while reversibility is an unattainable and unobservable ideal limiting behaviour (but useful for models)[14]. A quantum event once observed or detected is irreversible. In order for any event to be reversible it must be 'decoupled' from the rest of the Universe (and therefore be unobservable). This suggests a basis for 'understanding' apparent contradictions such as wave–particle dual behaviour: we must not expect unobservable objects to be 'real'.

It is important to note that very small (quantum) events as well as the macroscopic classical events fit the chaos model. Physicists have demonstrated this experimentally, while investigating the behaviour of a small number of iron atoms on a copper surface[15].

4.2 Wave States

Oscillations have the property that the displacement from a certain point is simply a function of **time**, and periodically passes through zero. For example, in the clock pendulum previously discussed we saw this in the displacement from the mid-point of the pendulum's swing. The property of waves that distinguishes them from oscillations is that there is a progression of the oscillation through **space**. However, the wave disturbance moves through the medium without displacing it permanently. Waves such as

- ripples initiated on the surface of a pond by a thrown stone

- the side-to-side disturbance of a periodically jerked rope or bowed violin string

- the propagation of sound waves through the air

cause no permanent shift of the water surface, rope, string or air although the wave transports energy.

Waves are a particular kind of structure[16]. This is most obvious in 'standing' waves, such as those in the air in organ pipes, where there is resonance between sound of a particular wavelength and a tube of a particular length. We characterise waves by their 'frequency' (the number of patterns repeated in unit time) and the 'wavelength' (the length of one repeating pattern, again demonstrated clearly in wind instruments). The wave velocity is the frequency multiplied by the wavelength, and depends on the type of wave.

If the **source** of the waves is moving towards us, each wave crest is closer to the one in front than if the source is stationary, so it reaches us sooner. Also, the frequency is higher. For a sound wave the pitch rises when the source and observer move towards each other. Conversely, it falls when the source is moving away from the observer. In everyday life this 'Doppler effect' or 'Doppler shift' (named after the nineteenth-century Austrian, Christian Doppler) is most noticeable in train whistles and aircraft engine noises.

Another of the important properties of waves is that they 'diffract': they spread out as they pass through an aperture or hole. If waves of a particular wavelength pass through a hole of a size comparable to that of the wavelength, there is 'interference'. As a result, there are alternating regions of

- reinforcement (where wave 'crests' coincide with 'crests', and 'troughs' coincide with 'troughs'), and of

- cancellation (where crests coincide with troughs of the wave)[17].

Our real life or classical view is that light travels in straight lines as we would expect massless particles to do. It is only when we get down to the interaction of light with small particles that we see these rules begin to fail, as with the interference when light passes through two small holes.

James Clerk Maxwell in the mid-nineteenth century proposed a model for light as an electromagnetic wave travelling through a vacuum (at a definite speed: the speed of light). His use of the term 'wave' implied the transmission of energy without the transfer of matter, analogous to waves on water or sound waves in air. Electromagnetic waves in this model do not even need the presence of matter for their transmission, unlike water waves or vibrating wires where we see a visible shape or sequence of waves in the medium. The wave model of light includes the Doppler effect, such as when an atom is moving relative to a light source. In a gas of atoms moving in random directions there is a 'Doppler broadening' of the spectrum[18]. We also include diffraction and interference in our light wave model. It 'explains' what happens when we shine monochromatic ('single-colour') light through a circular hole onto a screen and see a series of concentric bright and dark rings. The wave model of light is also useful in discussing partial reflection at surfaces and successive partial reflections in films of wavelength thickness. (The 'iridescence' generated by white light in the feathers of peacocks is an example.) In the case of glass, on average 4% of the visible light striking the air–glass interface at right angles is reflected. On meeting the second surface (the glass–air interface) the partial reflection varies from 16% to zero depending on the thickness of the glass and the 'colour' of the light. To account for partial reflection in the photon model of light we have to admit that we can calculate only the probability of a photon being reflected[19].

Should the wave model of electromagnetic radiation use transverse waves (like waves in a jerked rope) or longitudinal waves (like sound waves in air or the disturbance propagated through a coiled spring) (figure 4.1)? The phenomenon of 'polarisation' suggests that we should use a transverse wave model. Polarisation occurs when light passes through certain materials, such as tourmaline crystal or the commercial 'Polaroid®' sunglass lenses, where two layers of the material with opposite orientations completely extinguish transmitted light. The wave we generate by jerking from side to side the end of a rope lying on the ground is a transverse wave 'polarised' in the horizontal plane (the surface of the ground). Our model of electromagnetic waves is one of transverse oscillation. Unlike a rope lying on the ground, however, the wave is transverse in all planes through the direction of transmission. One 'polariser'

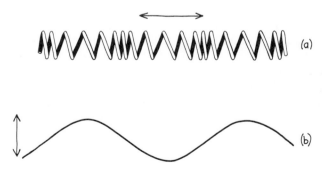

Figure 4.1 Typical waves: (a) a longitudinal wave in a coiled spring; (b) a transverse wave in a rope or musical instrument string.

transmits the wave motion in one particular plane, and if a subsequent polariser has its plane of polarisation at right angles to the first, no radiation can pass[20].

In laser light sources we can think of all the atoms as acting together, so the light is 'coherent' or 'in phase', in contrast to the situation in other light sources like tungsten lamps.

The wave model of light makes it easier to understand the nature of laser beams because of two properties. One is spectral purity. The light waves are of very similar wavelengths (and frequencies), unlike 'ordinary' light which is a combination of a range of wavelengths. The other property is spatial coherence: the waves are all 'in phase'. These properties enable us to focus laser beams to a spot nearly as small as a single wavelength (a few hundred nanometres for visible light). This focused light demonstrates by moving small particles that light carries momentum as well as energy, an idea easier to appreciate if we think of light as a particle or photon.

Louis de Broglie and Albert Einstein in the 1920s developed the idea that if 'particle' properties as well as 'wave' properties are associated with light, why should not 'wave' properties be associated with 'objects'? This should hold whether the objects are electrons, protons, atoms or macroscopic bodies. In this model all objects have an associated matter wave, characterised by a particular 'de Broglie wavelength'. Physicists have demonstrated experimentally the interference between atoms cooled to low temperatures. The distribution of atoms after passing through two slits shows the periodic intensity variations predicted for particles with those de Broglie wavelengths[21]. An interferometer can split an atom into two waves that then recombine and interfere, showing the need for wave-like behaviour as well as particle behaviour in our models of matter. Images of quantum mechanical standing waves of freely propagating electrons on the surface of a metal crystal have been obtained by scanning tunnelling microscopy[22].

We can use laser beams to control chemical reactions, and model the process as interference between matter waves when two different laser beams excite the molecules. At present the chemical industry varies the bulk properties of temperature and pressure to manipulate processes. Variation of intensities and relative phases of the laser beams has the potential to provide a much higher degree of control over the nature of chemical products[23].

It is helpful to think about the way in which a quantum model can describe other macroscopic processes. Consider the straight-line motion of an object from one point to another. We can model this as quantum behaviour with the particles of the object 'trying out' all possible paths simultaneously as if they were waves. Each of the infinite number of possible paths contributes equally to the total wave, but all paths 'cancel out' as a result of interference except the straight-line path and those paths close to it. Paul Davies[24] uses the 'popular-path' model of the routes taken by people strolling between two gates in a park to demonstrate this. Most people minimise their activity by taking a straight line, but a few take indirect routes. For macroscopic objects deviations from the straight line are negligible and the 'path of least action' or 'line of least resistance' corresponds to newtonian behaviour. However, for very small objects (quantum particles) like electrons these other paths are important. (In this 'popular-path' model we can think of overweight, heavy people being more likely to take the direct route than small, light people like children.)

Just as lasers provide coherent (in-phase) light waves that we can use to make optical holograms, we can generate coherent electron beams for electron holography and other electron interference effects. This is possible because the phase of an electron wave is sensitive to electromagnetic fields[25].

We should not confuse these wave-like characteristics of the **matter** wave model with those of the **electromagnetic** wave model. Matter waves are not emitted or radiated by the particle and do not travel at light speed, but they remain with the particle. Every particle has a wavelength, but macroscopic objects have wavelengths which are too small to be physically significant[26]. We are using the word 'wave' as a model: to describe something intuitively 'unnatural' by comparing it to something intuitively more 'natural', a similar physically perceivable phenomenon. The particles are not waves, any more than 'crime waves' are 'real' waves, but the particle has wave-like properties just as the incidence of crime spreads through a district in a wave-like manner[27].

It is a situation where

● phenomena originally seen as continuous (matter, electricity, light) are now described more precisely by quantised models,

● objects previously regarded as discrete are now better described by associated wave properties.

A quantum mechanical wave function, like a physical wave, has a probability amplitude (wave height or strength) and phase (extent of progress through a typical cycle), quantities that vary in space and time. In the probability model, the chance of finding the particle described by the wave function at a particular location is specified by the displacement. At a 'node' in the wave function, the probability is zero. We can explore the wave function and the nature of its interaction with fields within this comprehensive and powerful mathematical model.

With a model of a 'superspace' made up of a combination of all possible spaces, we may even describe the properties of space itself as wave-like, with interference effects[28].

4.3 Wave in a Box

It is a well known fact that a car radio does not receive radio waves in an underground tunnel. The waves interfere with their reflections from the walls and they cannot propagate through a space less than half the dimensions of the wavelength (many metres in the case of radio waves). We can visualise this as the shortest distance that allows a 'standing wave', a wave with a maximum amplitude at the middle and a minimum or node at the ends, like a plucked guitar string.

In a similar way, an antenna in a confined space cannot broadcast long wavelengths. An energetically excited atom that would otherwise radiate cannot do so if we confine it in a cavity. This leads to the technology of 'cavity quantum electrodynamics'[29]. (Quantum electrodynamics or QED is a quantum model for the interaction of electromagnetic radiation such as light with electrons and other charged particles[30].) The phenomenon of suppressed emission has been demonstrated with high-energy-state atoms having diameters of macroscopic size[31] as well as in solid state cavities. Although small cavities suppress the emission of photons, slightly larger ones matching the photon wavelength encourage emission. These are resonant cavities, comparable to an organ pipe providing a cavity in which a sound wave of particular wavelength can resonate. In such a system, we model an excited atom as oscillating between that high-energy state and a state made up of one photon and a lower-energy atom. This is analogous to the classical and easily observable system of two identical pendulums, linked by a weak spring, that exchange energy. One swings while the second is at rest, then the other swings with the first at rest. These effects provide the opportunity for a variety of novel devices, including the use of a weak laser and an optical cavity of mirror surfaces for the detection and counting of single atoms.

Lasers (using light) and masers (using microwave radiation) themselves depend on tuned cavity processes in which the wavelength of the energy transitions matches the cavity dimensions.

4.4 Heisenberg May Have Been Here![32]

One of the outcomes of the rapid development of quantum models in the 1920s was Werner Heisenberg's formulation of an 'uncertainty principle'. We cannot determine simultaneously to any desired precision both of the 'conjugate' or 'complementary' properties of a particle. We saw above that we cannot specify a particular path through space for a quantum particle. This follows automatically if a particle cannot have both a defined position and a defined motion, consistent with the model of the particle attempting to pursue all possible paths at once[33]. Wave–particle duality is one example of 'complementarity'. The wave properties and the particle properties of light and matter, observed in different circumstances, complement each other. Combining the wave and particle models provides a more complete description[34]. (We should note that in the future we may need to expand the models even further as our understanding improves.)

We can also visualise this uncertainty by considering what happens when we try to measure the position and velocity of a particle by shining a 'light' on it. Light (electromagnetic radiation of any kind) is quantised, so we must use at least one quantum of energy. If the particle is very small, one quantum of energy will be enough to change its velocity significantly—a 'Catch 22' situation: to observe something we must change it.

An example of a pair of conjugate properties is position and momentum. The more precisely we determine a quantum particle's position, the less certain we can be about its momentum, and measuring one produces uncertainty in the other. The classical model based on our everyday experience incorrectly assumes that position and momentum are independent: in fact they are complementary. The approximate, classical model is adequate for macroscopic objects, but not for quantum particles. The limit is a quantitative one: the uncertainty in momentum of a particle multiplied by its uncertainty in position can never be smaller than a certain quantity (defined by Planck's constant). It also follows from the Heisenberg uncertainty principle that electric and magnetic fields cannot both be zero anywhere. There is always a zero-point radiation field that we cannot remove ('vacuum field intrinsic fluctuations') occurring at all electromagnetic frequencies. This is another counterintuitive aspect of the quantum mechanical model[35].

The uncertainty principle imposes limits on the extent of our understanding and of the detail possible in any model of matter or of the Universe as a whole. One view of the quantum model is that a quantum particle has no 'real' position or momentum unless we measure them. If we measure its **momentum** it has no 'real' **position**, and the object exists with this 'real' momentum until we define its position[36].

Richard Feynman[37] interpreted the uncertainty principle as a warning about the limits we should place on our 'old-fashioned ideas' (inadequate models). Statement of this 'principle' would be unnecessary if we used more appropriate models in which these limits are 'built in'. It expresses our understanding that we cannot observe a **quantum** system sufficiently delicately to provide the amount of information we would consider essential for the complete description of a **macroscopic** system.

We can make use of this feature of quantum systems, which might appear to have only negative or inconvenient consequences. We can illustrate this by potential methods for 'private communication' or 'discreet communication' between two parties using public systems. (An example of private communication of this type is the use of coded messages between two people in newspaper classified advertisements for protection from third parties. Discreet communications are necessary where the two parties wish to protect themselves from each other, as in negotiations between opposing espionage organisations.) The definition of the quantum state of a photon requires not only its direction and frequency (energy) but also its polarisation: the direction of its associated electric field. We can determine this by the orientation of a polarising film:

- all light incident at right angles is transmitted when the polarisation is in a certain direction (probability of transmission unity)

- no light is transmitted perpendicular to that direction (probability of transmission zero).

Photons polarised half-way between these directions (diagonal polarisation) passing through an appropriate polarising filter 'forget' their polarisation. It is possible to use polarised photons in a device for discreet communications because if we know whether they have 'rectilinear' or 'diagonal' polarisation we can make observations on them to obtain further information. If, on the other hand, the wrong kind of measurement is made without this preliminary information, **all** information they are carrying is lost. Rectilinear polarisation and diagonal polarisation are 'complementary properties'. Attempts to measure one property result in randomisation of the other. These systems are now in the prototype stage of development[38].

One of the properties of electrons described in the 'spin' model is their 'orientation' relative to the direction of a magnetic field. Electrons with the same spin orientation are not likely to approach each other, but those with opposite orientations may do so. As described by the exclusion principle, electrons in atoms are 'paired', with the effects of an electron of one orientation 'cancelling out' the effects of an electron with the other orientation. Associated with an unpaired electron in an atom, ion or molecule are permanent magnetic properties, 'paramagnetism'[39].

4.5 Playing Cards and Four-Strokes

Our models of quantum particles are of two fundamental types, depending on the nature of their 'spin' or angular momentum, the way they interact with magnetic fields and their group behaviour.

We do not believe particles have axes about which they rotate, so it is misleading to think of particles as little tops spinning about their axes. Rather, we should think of their symmetry properties: what they look like from different directions. Other ways of asking this are: 'How far must they be rotated to look the same as they did originally? Under what conditions does their appearance remain the same? In what situations are their properties conserved?'.

- A spin-zero particle looks the same from every direction, no matter how far we rotate it, like an unpatterned dinner plate in two dimensions, or a perfect sphere in three dimensions.

- A spin-one particle on rotation looks the same once it has turned through a complete revolution (like a clock face).

- A spin-two particle looks identical twice in every rotation, like a playing card rotated in its plane. This is an integral multiple of spin or rotation property, and is a model that is easy to picture.

- The other possibility is not quite as familiar. Some particles have a spin of one-half, and one must turn them through two complete revolutions before they look the same. But we can even model this behaviour in the real world. In a four-stroke internal combustion petrol engine one combustion 'cycle' requires **two** complete revolutions of the flywheel. Another model is a 12-hour clock face representing 24 hours, showing the same time only after **two** revolutions.

(In 1928 Paul Dirac devised a mathematical model for the spin-one-half situation, consistent with both quantum mechanics and special relativity[40].)

Electrons, protons and neutrons as well as some nuclei and atoms (such as carbon-13 and chlorine-35 nuclei) have **half-integral** spin or rotation properties. They have group behaviour that follows 'Fermi–Dirac statistics', and are 'fermions', named after Enrico Fermi. Fermions obey an exclusion principle formulated by Wolfgang Pauli in 1925. At any given time each particular quantum state can be occupied by only one of this type of particle, so it is not possible for all particles to cluster in the lowest energy state. Fermi–Dirac statistics apply to electrons and other 'conserved' particles. Conserved particles are those particles whose number in the Universe always stays the same[41].

On the other hand, photons have unit spin and belong to the 'boson' class of particles with **integral** spin or rotation properties. They obey 'Bose–Einstein statistics' (named after Satyendra Bose and Albert Einstein). They are not subject to the exclusion restriction. This means that as the system approaches the absolute zero of temperature, all bosons in a given system can condense into a single quantum state, the one with lowest energy. Bose–Einstein statistics apply to particles which are not conserved, such as photons (whose numbers in the Universe are always changing).

The wave function describing a group of bosons is unchanged when two bosons change places. In contrast, the sign (positive or negative) of the wave function of a group of fermions reverses when two fermions are exchanged.

The particles constituting atoms (electrons, protons, neutrons) are fermions, but when they are combined the new atomic particle they form is either

● a boson, if made up of an even number of fermions, like helium-4 (two electrons, two protons, two neutrons), or

● a fermion, with an odd number of fermions, like helium-3 (two electrons, two protons, but only one neutron)[42].

The spin concept helps us to rationalise the natures of the interactions between particles:

● Fields associated with odd-integer spin particles can produce both attractive and repulsive interactions. In electromagnetics, the photon, with spin one, carries the force. The interaction is attractive between oppositely charged particles and repulsive between like-charged particles.

● Fields of even-integer spin particles (which include the spin-two graviton associated with gravitational fields) produce a purely attractive interaction[43].

4.6 Quantum Leap

In discussing the properties of matter the expression 'quantum leap' is usually inappropriate because quantum behaviour is relevant to **small** quantities, and classical descriptions are adequate for macroscopic bodies. However, there is an important situation where this is not so.

At low temperatures, helium is a 'quantum liquid': classical mechanics is inadequate to describe not only its atomic properties but also its bulk behaviour. This is a very clear demonstration of what happens in a less obvious way in all macroscopic materials. The energy states of macroscopic as well as of atomic objects are discrete (although so closely spaced that we can treat them as continuous in most cases). Low-temperature helium at low pressures remains liquid because the positions of these light atoms remain uncertain for reasons described by the Heisenberg uncertainty principle. They have too much motion even at these low temperatures to be able to form a solid. In most fluids, collisions between particles cause shifts between close energy states resulting in energy dissipation. In contrast, low-velocity collisions do not raise ground state (superfluid) liquid helium into a higher state in this way, so energy is not dissipated. We can appreciate the particular properties of superfluid helium only by considering its quantum mechanical properties. Helium-4 atoms (made up of two protons, two neutrons, two electrons) are bosons. At very low temperatures a single quantum state describes the behaviour of not only the individual particles but also the whole macroscopic system of liquid helium[44].

Superconductors also have low-temperature behaviour that we can model in terms of quantum effects being important in macroscopic as well as molecular properties[45]. We can observe the processes associated with quantum mechanical 'tunnelling' in a carefully devised experimental system, in particular the potential energy of a magnetic flux associated with a superconductor[46]. In a recent development, experiments have shown that tunnelling can occur between different magnetic states in **macroscopic** crystals of manganese acetate: the scientific frontier where quantum physics (microscopic and quantum mechanical) meets classical physics (macroscopic and classical)[47].

In summary, our approach is to:

- Accept the model of a 'quantum world', fundamentally different from the classical world.

- Simplify it for most systems by using classical models as the limiting case.

- Retain the quantum model for a few 'special' macroscopic systems.

The relativity model says that scientific principles should be the same for all observers, regardless of their speed. The extent of validity of this statement surpasses Newton's model ('the laws of motion') to include speeds approaching that of light. When an observer travels at a higher speed there is an increase in the observer's mass, with the result that incoming light (as measured by the observer) always has the same speed. Speed, distance, time, energy and mass are interconnected[48].

One of the consequences is the equivalence of mass (m) and energy (E) expressed by the famous equation $E = mc^2$, the quantity c being the speed of light. Any object possesses a 'rest mass' and a 'rest energy' (the energy equivalent of its rest mass). But it may have additional energy and 'relativistic mass' as a result of its speed. When the speed of a particle approaches the speed of light and the particle's kinetic energy becomes comparable to its rest energy, relativistic effects become important. For this reason, we often express the masses of the fundamental particles in energy units.

The link between the quantities we choose to call 'mass' and 'energy' demonstrates that our models incorporate only some aspects of a more complex reality. These model quantities are important (and at the moment essential) to our 'understanding' and manipulation of matter, but they are not necessarily the most fundamental quantities.

For such 'relativistic particles' (which include electrons near an atomic nucleus, with speeds significant compared with light speed) the number of quantum states does not depend only on properties like its rest mass. The number also increases with an increasing ratio of total energy to rest energy. As the relativistic mass increases, the number of quantum states increases, and quantum mechanical particles behave more like classical macroscopic objects, with nearly continuous quantum states available. It is interesting that two 'unusual' or unfamiliar models (quantum states and relativity) are here interacting to result in more classical or 'usual' model behaviour at an atomic scale[49].

Endnotes

[1]Further reading Kaye (1993, Chapter 1), Home and Gribbin (1994).

[2]Home and Gribbin (1994).

[3]See **8.1 Atom Models**.

[4]Further reading Baggott (1990, 1992), Horgan (1992), Rae (1986), Englert *et al* (1994), Bernstein (1996). In the terminology of Thomas Kuhn (see **1.5 Hindsight and Foresight**) alternative paradigms are incommensurable—they are not contradictory or incompatible.

[5]See **4.2 Wave States**.

[6]Aspects of the approach of Rupert Sheldrake (**15.13 Evolving Universes**), that suggest a collective memory in the universe associated with the spread and expansion of ideas, have something in common with this behaviour.

[7]Further reading Home and Gribbin (1994). See **4.5 Playing Cards and Four-Strokes**.

[8]See **7.5 Tunnelling**. Further reading Chiao *et al* (1993). The barrier is intact despite the tunnelling, and the quantum particle has no real velocity as it passes through the barrier. There are also interesting implications in the behaviour of quantum particles concerning the proposition that nothing travels faster than the speed of light.

[9]More strictly, as the relevant parameters of the system become large compared with Planck's constant, h, 6.63×10^{-34} J s. See also **2.6 Limits and Dimensions**.

[10]Described in **6.2 Probing Matter** and **8.10 Surface Images**.

[11]Discussed in **4.6 Quantum Leap** and **11.3 Superfluid Helium**.

[12]The questions of quantisation of electric charge and of mass are discussed in **7.3 Charge Integrity** and **8.1 Atom Models**. The conductance of a one-dimensional quantum conductor is also quantised: **7.6 Delocalisation**.

[13]Further reading Heller (1996), Wilkinson *et al* (1996), Gutzwiller (1990, 1992), Uzer *et al* (1991), Shimony (1988), Rae (1986).

[14]Other examples of limiting behaviour can be found in **3.2 Dynamics**, **4.1 Ultraviolet Catastrophe**, **5.3 Absolute Zero** and **9.5 Classifying Matter**.

[15]Further reading Heller *et al* (1994), Heller (1996). See **8.4 Quantum Chaos**.

[16]Waves are examples of dissipative structures: **5.7 Ordering**.

[17]See also **6.3 Transformation**.

[18]There is also another frequency shift due to relativity, **4.6 Quantum Leap**. Further reading Phillips and Metcalf (1987).

[19]Further reading Feynman (1985, Chapter 1).

[20]The rotation of the plane of polarised light by its interaction with matter is considered further in **9.12 Handedness**.

[21]See also **8.8 Cooling Atoms**. The de Broglie wavelength was found to be 0.06 nanometres for helium atoms and 0.02 nanometres for sodium atoms. Further reading Hecht (1991a), Chu (1992, p 53).

[22]See **8.10 Surface Images**. Further reading Zettl (1993).

[23]Further reading Brumer and Shapiro (1995).

[24]Davies (1982).

[25]These effects have been used for imaging quantised flux lines in superconductors (**13.7 Superconductors**). Further reading Ball and Garwin (1992, p 764), Bishop (1993).

[26]The wavelength of a matter wave is given by $h/(mv)$, where h is Planck's constant, m is the mass of the particle and v is its speed. A mass of 100 grams with a speed of 90 kilometres per hour has a wavelength of about 10^{-32} cm.

[27]Davies (1980, Chapter 3).

[28]Further reading Davies (1980, Chapter 5).

[29]Further reading Haroche and Raimond (1993), Feynman (1985). The properties of electrons can be described in a similar way: see **7.4 Electron in a Box** and **7.7 Quantum Dots**.

[30]Further reading Home and Gribbin (1994), Craig (1992). Chapter 6 deals with the inter-relationship of energy and matter.

[31]See **8.3 Rydberg Atom**.

[32]Quoted by Weber (1992, p 280).

[33]Further reading Davies (1985, Chapters 2, 14).

[34]See **4.1 Ultraviolet Catastrophe**. Further reading Englert *et al* (1994).

[35]The use of the zero-point radiation field in describing intermolecular processes is described in **10.1 Molecular Interactions**.

[36]Rae (1986, Chapter 4). Further reading Hawking (1988).

[37]Feynman (1985).

[38]Further reading Bennett (1992), Bennett *et al* (1992), Horgan (1992), Collins (1992). It is also possible to use pairs of mutually inverse transformations to encode information for these public key cryptography purposes, but the mathematical procedures are complex: see also **6.3 Transformation**.

[39]An ion is an atom carrying a positive or negative charge. Ferromagnetism, a particular type of magnetic property associated with ordered structures in some solids, is discussed in **13.9 Magnetic Matter**.

[40]Further reading Hawking (1988, p 67), Postle (1976, Chapter 10). See also **4.6 Quantum Leap**.

[41]Discussed further in **7.8 Colourful Quarks**, and the properties of helium-3, which is also a fermion, are discussed in **11.3 Superfluid Helium**.

[42]This distinction is important when we describe helium superfluidity (**11.3 Superfluid Helium**) and superconductors (**13.7 Superconductors**).

[43]Further reading Goldman *et al* (1988), Hawking (1988).

[44]See also **11.3 Superfluid Helium**.

[45]See **13.7 Superconductors**. Further reading Bardeen (1990), Lounasmaa and Pickett (1990).

[46]See **7.5 Tunnelling**. Further reading Shimony (1988).

[47]Further reading Stamp (1996), Thomas *et al* (1996).

[48]See **13.7 Superconductors**. Further reading Postle (1976, Chapter 5).

[49]Further reading Hawking (1988), Norrby (1991), Goldman *et al* (1988).

CHAPTER 5

THERMODYNAMIC MATTER

We now come to 'thermodynamics', probably the most important model in science. Some scientists would say that this model best describes the behaviour of materials. Central to the thermodynamic model is 'energy', and although we can define energy broadly as the capacity to do work, it is a very elusive concept. The 'energy well' model provides a useful introduction to energy and to thermodynamics.

5.1 Energy Well Model

Consider a well, a brick-lined hole in the ground into which rainwater finds its way and from which we can obtain water only by expending energy by pulling up a bucket or using a pump. This models a 'gravitational well', the point of lowest practical gravitational potential energy for an object because of the attraction between the mass of the object and the mass of the Earth. We can generalise this model of an energy well to encompass all forms of potential energy, including interatomic potential energy. For quantum particles the possible energies are restricted to certain levels within the well, as though there are protruding bricks at irregular intervals in the sides of the well where we can rest a bucket of water while bringing it up from its 'ground state'.

Rather than using as our model a conventional well as a long, cylindrical hole in the ground lined with bricks, I like to consider potential energy wells as looking like irregular ponds or lakes. They have varying depths, which we can use to portray the energy as a function of two or more distances in a system. In order to make our model a little more realistic, we can visualise a water storage lake with a level that varies as water flows in or is drawn off (figure 5.1). Jetties are built at various levels to accommodate boats on the changing levels. There is only one position for the water level convenient for the use of each jetty, so we control the water drawn off to maintain the level at one of the jetty positions. This model conforms to the potential energy requirement that lower levels must be filled first.

Besides the energy minimisation model, there is a related model based on balancing forces. In the interaction between a pair of atoms, there is a balance between the attraction and repulsion. The position of this balance determines what we describe as the atomic diameter, the typical separation of colliding atoms. Objects

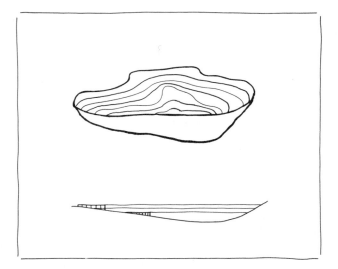

Figure 5.1 A potential well can be visualised as a water storage lake with a level that varies as groundwater runs in or water is drawn off. Jetties can be built at various levels to accommodate boats on changing levels, but there is only one water level that is convenient for the use of each jetty.

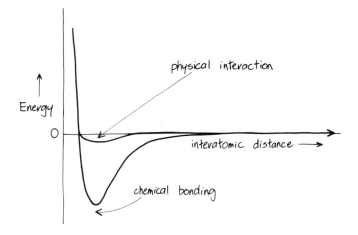

Figure 5.2 Energy wells for physical and chemical interactions.

held together by strong attractive (or cohesive) interactions tend to have lower energies (because one has to use more energy to separate them).

Now apply the energy well model to two interacting atoms (figure 5.2). The potential energy surface representing these two atoms has a constant, 'flat' value when the atoms are far apart. By convention, we take the energy to be zero at infinite separation, and it becomes negative as the atoms interact. The energy decrease becomes steeper due to attraction when they are closer, then it reaches a minimum before steeply rising with strong repulsion at interatomic distances much less than atomic diameters.

- If the 'attractive' or 'cohesive' part of the potential energy surface is deep, the two atoms undergo a 'chemical interaction' to give a 'chemical bond' that may be very long-lived.

- If it is shallow there is only 'physical interaction' between the atoms, and a brief encounter.

As with all models, this classification into 'chemical' and 'physical' models is too sharp: in real life there are also intermediate situations.

It is important to note that these energy wells are often very shallow relative to the total magnitude of the energy, just as water will always find its way to the lowest level, no matter how small the difference in depths. Frequently when we calculate energies we are looking for very small differences between very large opposing effects, so the results are difficult to predict accurately[1]†.

As an alternative and more picturesque analogy, we can visualise the environment of particles in matter as a landscape made up of hills, ridges and valleys. Climbing out of a valley onto a ridge takes energy, just as it would in a real landscape, where water would move to the lowest available level as in the energy well analogy. We are more familiar with the description of geographical surfaces on land rather than under water, so potential energy surface models[2] usually use terms such as valleys, mountains, mountain passes and plateaus. They are variations of energy well models.

5.2 Thermodynamic Model

'Thermodynamics' is a word made up of two parts meaning 'heat' and 'work'. This model enables us to discuss the transformations of energy in a system in much greater detail than when we use the simpler energy well model. Like the energy well model, it is subject to certain limitations on the rates at which processes occur (as I explain below). Engineers originally developed the thermodynamic model to help them improve the efficiency of steam engines[3]. Thermodynamics is a model that we can apply to any system (without even knowing its particular structure) and it is widely used by chemists and physicists as well as by engineers.

First we must define the 'system', carefully separated from its 'surroundings'[4]. When **matter** can be transferred across the

† Endnotes for this chapter can be found on page 122.

boundary, we say that the system is system is 'open', otherwise it is 'closed'. If **neither energy nor matter** can cross the boundary of a closed system we say it is 'isolated'. We call the total energy of an isolated system its 'internal energy', because it is internal to the system. This is a state property of the system depending only on its current condition and not on how it got there.

An important idea built into the thermodynamic model is that of 'reversibility', a word with a much more precise meaning in thermodynamics than in everyday use. A reversible change is not just a change that we can reverse by a change in conditions, but one that can be reversed by the smallest possible ('infinitesimal') change in conditions. A system controlled by such a reversible process is 'at equilibrium'. This is a dynamic concept, and if we wish we can visualise innumerable individual, molecular interactions that have the effect of evening out any imbalance in energy or concentration anywhere in the system. The strong mixing effect rapidly achieves this situation of

- microscopic or molecular disorder, at the same time as
- macroscopically reproducible equilibrium and order[5].

Equilibrium is an essential aspect of the thermodynamic model. A system is in equilibrium with its surroundings if negligibly small changes to the conditions (in either direction) result in changes in its state. For example, if you have a mixture of ice and water at zero degrees Celsius (the 'freezing point' of water), the addition of even a small amount of additional heat causes a little of the ice to melt. The removal of heat causes some of the water to freeze. Ice and water are in thermal equilibrium and form a reversible system at this temperature.

In practice, very few real life systems are ever at equilibrium, and no systems converting energy into useful work are ever at equilibrium[6]. Nevertheless, this model of an ideal process has proved very useful in describing limiting behaviour[7].

The concept of 'temperature'[8] is closely associated with the idea of thermal equilibrium. Energy flows from the body at the higher temperature to the one at the lower temperature. This occurs until the bodies are at the same temperature and thermal equilibrium exists. Because the concept of 'temperature' depends on this dynamic process, we cannot apply the term to a single molecular particle.

The well known statement in classical mechanics that 'energy is always conserved' is more correctly expressed: 'the **internal energy** of an **isolated** system is constant'. It conforms to the observation that it is impossible to construct a 'perpetual motion' machine (a device that will continue to run indefinitely without an external energy input). We can distinguish between the processes of

- transfer of energy using random thermal disordered motion or heat, and

- carrying out ordered motion or work.

Both are methods of transferring energy. James Prescott Joule initiated a statement summarising some important aspects of the thermodynamic model . The 'first law of thermodynamics' says that

- in the combination of processes involving **heat flow** and **work done**, energy is conserved.

The internal energy of a system is constant unless changed by the performance of work or heating. It is independent of the method of preparation of the system or the method of attaining the energy.

(An obsolete model that we abandoned rather than integrating into newer models is that of heat as the massless fluid substance 'caloric'. Sadi Carnot believed that the quantity of caloric or heat was conserved while work was being done. This was a 'misleading model'[9], although other contributions made by Carnot laid the foundations for thermodynamics. We should regard work and heat not as 'substances' but as ways of transferring energy, and energy itself is not a 'substance'[10].)

When scientists investigated situations in particle physics where it appeared that energy was not conserved they were led to the 'discovery' of new fundamental particles. (That is, they made their models of matter more complex and detailed to accommodate some of the discrepancies between previous models and observations of reality.) Just as important has been the use of the energy conservation principle in the chemistry and physics of materials at the everyday or macroscopic level.

'Work' can take various forms, but we often restrict it to situations where a force causes an object to move or otherwise respond to the force. In chemical thermodynamics we can further restrict it to the processes associated with expansion (and contraction) against the surrounding atmospheric pressure. The work associated with the production of gases in chemical reactions is particularly important. This version of the model therefore excludes other forms of work, such as electrical work involved in electrochemical reactions, as well as expansion of surface area (which also requires work to be done)[11].

When a chemical reaction or other change of state occurs at constant volume (for example in a strong pressure vessel) the change in internal energy is equal to the amount of heat absorbed or liberated. There is no expansion work against the surroundings because the system is isolated from these effects. A good example of this is the 'bomb calorimeter' that technologists use to measure the calorific value of foods. The amount of heat evolved equals the internal energy change associated with the combustion or burning of food, to provide the figure stated on the packets we buy as diet-conscious consumers[12].

We do not live in a world where processes take place at constant volume (or at constant temperature). Rather, in our world the natural atmospheric pressure is what is approximately constant, so it is reasonable that the thermodynamic model emphasises processes that take place at constant pressure. (We can speculate that a hypothetical dolphin civilisation, in a marine environment that has approximately constant temperature but considerable variation of pressure depending on depth, would have developed a rather different set of models.)

When a system increases its volume (for example while producing gas in a chemical reaction, or by evaporating) against a constant external pressure, it does work to drive back its surroundings. This means that the heat evolved or absorbed is no longer a measure of the change in internal energy (the change in the total energy of the system) as it would be at constant volume. The surroundings are now involved. However, because we can measure it so conveniently and because it is so useful, we give this energy quantity (heat change at constant pressure) a special name: the 'enthalpy change'.

The thermodynamic model provides helpful practical information linking the number of chemical species and the number of distinct phases under specified conditions. This is summarised in the so-called 'phase rule'. However, like all models, the assumptions built into this model can cause limitations. One of these is the existence of what we may call 'nanoemulsions'. These are nanometre-sized groups of molecules, smaller than those in surfactant 'micelles' or 'microemulsions'[13] and not detectable by microscopy. Such a system looks as if it is homogeneous or single phase, so the phase rule model gives apparently inconsistent results[14].

5.3 Absolute Zero

Temperature is the most important aspect of the thermodynamic model. As I have shown already, the quantity we call 'temperature' is intimately associated with the idea of thermal equilibrium. Although we may think we are familiar with temperature because of our extensive use of thermometers, it has subtle aspects that are not immediately obvious. For example, there is a definite lower limit to its value (zero on the absolute temperature or kelvin scale) and this limit is unattainable (the 'third law of thermodynamics'). This is perhaps the most fundamental example of 'limiting behaviour'[15]. P W Atkins[16], in his entertaining account of 'The Second Law', has summarised the three 'laws' of thermodynamics in an alternative form.

- Heat can be converted into work, but

- completely only at absolute zero, and

- absolute zero is unattainable!

It is unreasonable to expect an answer to the question of **why** we cannot reach a temperature of absolute zero. The most honest response is that this is one of the aspects of the thermodynamic model that is consistent with all our observations, and we accept the model because it works[17].

There is an interesting conceptual model that tells us something about temperature. Maxwell invented a hypothetical molecular-sized 'demon' with a very special (and we believe unattainable) skill. Maxwell imagined two containers of gas at the same temperature separated from each other by a 'door' of molecular size. This door was controlled by his 'demon'. The demon let only the high-speed (high-energy) molecules move from left to right through the door, and only the slow molecules cross in the opposite direction. The demon itself required no energy input. The result of this hypothetical activity was to cause the temperature of the gas on the right to rise, and on the left to fall. This behaviour is the opposite of all our observations that show heat flow always causing temperature differences to decrease rather than increase. We do not believe that any device like this is possible, and the Maxwell demon is a 'negative model', clarifying an aspect of the behaviour of matter by illustrating the opposite.

5.4 Energy and Entropy

So far, I have suggested that the only criterion for stability of a system (and the driving force for change) is the minimisation of energy. There is an even more fundamental foundation of the thermodynamic model: the drive towards disorder. The quantity used as a measure of the degree of disorder or randomness is 'entropy'. We can approach entropy from either the thermodynamic model or the 'statistical thermodynamics' model. Ludwig Boltzmann considered matter from the point of view of atoms, and related the entropy to the natural logarithm of probability multiplied by a constant factor, the 'Boltzmann constant'[18]. From a historical point of view, the first model used (by Sadi Carnot, in the early nineteenth century) in the development of this concept was the steam engine[19].

We can approach entropy also from the direction of 'information science' or 'information theory', the science created by Claude Shannon in the middle of the twentieth century[20]. That the quantitative model of thermodynamics has perspectives as diverse as steam engines, molecules and information science testifies to its fundamental importance.

Although the quantity of energy is conserved in any process, the way in which that energy is distributed is not reversible:

- There is a general movement of the total quantity of energy in the Universe from 'hot' to 'cold' rather than the reverse.

- There is a lack of symmetry in the flow of energy, just as there is in the flow of time.

The 'second law of thermodynamics' describes this 'dissymmetry' or 'irreversibility' of nature in terms of entropy. One way of more readily understanding the reason for spontaneous increase in entropy is by asking why disordered conditions are more probable (more favoured or more desirable) than ordered ones. We should also ask what we mean by 'order'. Disorder is more probable than order because there are many more ways of disordering things than of ordering them. There are many more states or configurations accessible in the disordered condition. As an illustration, if we notice that the books on a library shelf are in the sequence of their call numbers (the ordering of the books repeating the arrangement in the catalogue file) we assume that they were arranged that way by library staff. We are confident that this order could not have occurred by accident because there is a very much larger number of disorderly sequences, far more than the total number of books. (Of course, when it comes to systems occupying a multiplicity of states at the molecular level, from our macroscopic perspective they all look the same: hence the need for mathematical models.)

The numbers of atomic particles in materials are so much greater than the number of books in the library that the possible number of disorderly arrangements is overwhelmingly larger than the number of ordered arrangements. We can therefore say with confidence that natural processes always move towards an increase in disorder, and that spontaneous decreases in entropy (decreases in disorder) are impossible in isolated systems.

- A decrease of entropy in a system is possible only when accompanied by an increase of entropy in its surroundings.

The formal statement of these observations is the 'second law of thermodynamics'.

The thermodynamic model involves the definition of entropy as the quantity of heat divided by the temperature, so the lower the temperature the less useful the heat and the greater the entropy.

Whenever heat flows, entropy increases. Although it is a less familiar idea than that of heat, and our bodies cannot detect it as easily, entropy as part of the thermodynamic model is no more and no less real than heat[21].

The thermodynamic model emphasises the importance of the internal energy of a system. In addition, thermodynamics uses entropy as a measure of disorder such that a system always arranges itself to reach a balance between minimising its energy (at constant pressure, strictly its enthalpy) and maximising its entropy. The relative importance of the two factors in determining the position of equilibrium depends on the temperature, with the influence of entropy increasing as the temperature rises. At relatively low temperatures the changing entropy of a system has less effect on the position of equilibrium, so minimisation of energy dominates. At higher temperatures changing the entropy has a more marked effect, so maximisation of entropy dominates.

The thermodynamic model relates

- **energy** changes taking place in a **system**, to

- associated **entropy** changes occurring in its **surroundings**.

In the thermodynamic model, entropy increase is the driving force for all natural change. There is a natural tendency towards greater entropy or disorder in the **combination** of system plus surroundings, that is, in the Universe as a whole. For changes of state there is always a drive that favours gases over solids and liquids, regardless of the temperature. Water evaporates even on a cold day and snow 'sublimes' (evaporates from the solid state) even at the lowest temperatures. However, for most purposes it is more convenient to focus on the **system** by using the 'minimum energy, maximum entropy' model. We should not feel any reluctance to use whichever model is of the greatest value to us or most convenient in the particular circumstances. It does not matter if the model we choose does not happen to be the most fundamental, or complete, or sophisticated model, as long as it helps us to understand or to achieve some goal[22].

It is helpful to think about Maxwell's demon (introduced in the previous section) in terms of entropy. Superficially, there is a decrease in entropy (increase in order) because the demon has separated fast molecules from slow ones. However, to see the molecules, the demon needs to illuminate them with radiation different from that in the rest of the system, and the light source operates irreversibly. We can show that the entropy created by the light source always exceeds the entropy lost by sorting the molecules, so the 'second law' is not invalidated even in this hypothetical situation[23].

5.5 Death and Taxes

In the development of our civilisation, Nature's dissymmetry associated with the natural direction of change resulted in heat being used in technological processes before work was obtained from heat. So we were able to burn fuels for heat long before we used a steam engine to do work for us. It is much easier to light a fire than to build an engine that uses the energy of fuels as work. This apparent inequity in nature, conveniently modelled by entropy, has a parallel in the apparent inequity in taxation.

Most spontaneous processes in nature occur by systems releasing some of their internal energy to the surroundings as heat: they are 'exothermic'. However, not all this heat is necessarily available to do work. It is helpful to explore the process of obtaining useful work from a system that can undergo spontaneous reduction in its internal energy, such as producing steam pressure from the chemical processes of burning coal. If the entropy of a system **decreases** as its internal energy decreases (that is, a 'less favourable' change occurs), not all of that internal energy is available for doing external work. It is as if nature is demanding a tax on internal energy as it converts energy into useful work. The tax required is energy as heat for the generation of sufficient entropy or disorder in the **surroundings** to balance any entropy loss in the **system**. This idea leads to the name 'free energy' for that part of the internal energy available for us to use as work.

Sometimes we are fortunate enough that the entropy of a system **increases** at the same time as the internal energy decreases. We can get an amount of work out of the process that exceeds the internal energy change. In this particular situation it is as if nature is giving us a tax refund. There is a dissymmetry: although both heat and work are methods of transferring energy, nature taxes the process of converting heat into work, but does not tax the dissipation of work as heat[24].

In using this 'tax' model, we should remember that our individual tax refund is not greater than the amount of tax paid. For the community as a whole the total tax paid is greater than the total refunds: tax revenue is always positive. The same is true of entropy.

Some societies use 'negative income tax' for social welfare purposes, and in much the same way there are some 'endothermic' (heat-absorbing) processes that are spontaneous. As the system changes, the surroundings become colder. If the entropy increase in a spontaneous process is sufficiently great, this can drive the process. Heat flows into the system from the surroundings: a net tax 'refund' for the system paid for by the rest of the 'community' (the surroundings).

We can expand the tax model to encompass the improbable phenomenon of life. James Lovelock[25] has compared it to the process carried out by a skilled accountant, 'never evading the payment of a required tax but also never missing a loophole.'

There are other models we can use to illustrate the thermodynamics of processes. A *Journal of Chemical Education* 'Tested Demonstration' describes an easily constructed physical model[26].

5.6 Cussedness Model

We can sum up many results of the thermodynamic model by one very simple idea. A system at equilibrium responds to any attempt at disturbance by trying to minimise the effects of that disturbance. Formally, this is called 'Le Chatelier's Principle', but I like 'cussedness model' as an appropriate descriptive title:

- When we apply pressure to a system at equilibrium, the system adjusts by trying to reduce its volume and so minimising the pressure increase. (A chemical reaction at equilibrium does this by favouring the direction that reduces the number of molecules.)

- When we raise the temperature, the system tends to absorb thermal energy from the surroundings and convert it to internal energy (the process shifts in the endothermic direction).

- When we lower the temperature, the system opposes the reduction in temperature by an exothermic shift that releases heat to the surroundings.

5.7 Ordering

Different states of matter have different degrees of disorder or entropy:

- ice crystal is more ordered than liquid water,

- magnetised iron has atoms aligned in a more orderly way than unmagnetised iron,

- two liquids possess less order when mixed than when separate[27].

When we cool a material, the higher-temperature phase becomes unstable relative to the more ordered phase, and at some temperature a 'phase change' or 'phase transition' usually occurs.

As well as 'permanent' ordered structures which we recognise readily, in both molecular and macroscopic materials, 'temporary' ordering also occurs. There are transient systems which are relatively low in entropy (they are quite 'ordered') but which exist only where there is sufficient free energy flow. Structures showing stationary or cyclic patterns or travelling waves in the midst of a dynamic or changing system appear in these transient systems.

Examples occur in structures as simple and short-lived as growing piles of sand and the eddies and plumes in 'convective turbulence'[28], but they occur also in systems as complex and (relatively) enduring as life itself. These are 'dissipative structures':

- they have structure without the permanency of solids,

- they maintain decreased internal entropy at the expense of entropy increase in the environment, and

- they dissipate and disappear when their energy sources are exhausted.

Such structures are fluctuations allowing systems to achieve locally states that are far from thermodynamic equilibrium. Kauffman[29] quotes: 'Life exists at the edge of chaos', and we can make a similar statement about all dissipative structures[30].

Dissipative structures can survive more readily if they are associated with **solid** structures. At the simple level, we constrain eddies in the air to provide musical notes in the resonant hollow tubes of musical instruments. (Chemists used these 'standing waves' as models for electron 'waves' around nuclei of atoms.) Life maintains its infinitely more complex dissipative structure by chemical information in DNA molecules[31].

The relatively simple structures based on interactions between chemical reactions and the diffusion of reaction intermediates or products such as the patterns named after Alan Turing[32] have received considerable attention. Our heartbeat is a periodic process resulting from a complex array of chemical reactions and physical transport processes in our bodies. There are simple chemical examples in the laboratory, and these can serve as models of more complex behaviour, with striations or coloured bands that resemble those on the skins of animals.

There also exist spatially localised excitations ('oscillons') which can be considered as structural units that tend to assemble into larger structures, just as molecules assemble into atoms. These dissipative systems occur in systems like viscous fluids and granular materials[33].

5.8 Phase Changes

We can obtain useful information by applying the thermodynamic model to property changes associated with phase transitions or 'changes of state', and for this purpose phase diagrams are important aids[34]. In the description of such processes involving four very common states of matter, we use pairs of specialised terms to describe the forward and reverse changes. Most are in everyday use, but a few may be unfamiliar:

- **Vapour/liquid**: 'condensation' or 'liquefaction' in the forward direction; 'evaporation' or 'vaporisation' in the reverse; 'boiling' when the vapour pressure equals atmospheric pressure.

- **Liquid/crystal**: 'freezing' or 'crystallisation' in the forward direction; 'melting' or 'fusion' in the reverse.

- **Liquid/glass**: 'vitrification' in the forward direction; 'softening' in the reverse.

- **Glass/crystal**: 'crystallisation' or 'devitrification' in the forward direction[35].

- **Vapour/crystal**: 'deposition' in the forward direction; 'sublimation' in the reverse.

Volume changes accompany phase changes (for example, the volume of water increases when it freezes, so ice floats in water). There is also a change of enthalpy (the heat change at constant pressure: the heat provided to ice to make it melt). This used to be called 'latent' (hidden) heat, in contrast to 'sensible' heat (detectable by our senses). Boiling water stays at the same temperature while heat is being supplied to it because vaporisation requires considerable energy. Because the enthalpy of the water changes without an accompanying temperature change, we call such phase transitions 'first-order transitions'.

In contrast, order–disorder transitions in alloys[36], ferromagnetism transitions and transitions between fluid and superfluid phases in helium do not have abrupt changes of volume and enthalpy at the transition temperature. We call these 'second-order transitions'[37]. In this case the transition occurs gradually as the temperature approaches the transition point from below, becoming complete at the transition temperature.

In either first- or second-order transitions, another phenomenon may intervene. The higher-temperature phase may persist below the transition temperature in a 'metastable' state after cooling even though in the thermodynamic model it is not the preferred phase. This is 'supercooling' or 'undercooling'. The metastable state is temporarily stable as long as disturbances are not too great. Solid–solid phase transitions under the influence of changing temperature and pressure (such as some of those in the Earth's crustal rocks) are particularly prone to metastability.

Occasionally we notice rapid phase transitions of metastable systems accompanied by shock waves. Because the enthalpy changes in phase transformations are small compared with chemical reactions these phase transitions do not usually cause much damage. Chemical explosions provide 'models' for these rapid phase transitions. In 'low' or 'deflagrating' chemical explosives the shock front is subsonic, while in the 'high' or detonating explosives the reaction propagates behind a shock front at a supersonic rate. I have suggested previously[38] that rather than being analogous to the low chemical explosives, there are phase transitions that may occur at supersonic rates. If this is so, large volumes of metastable material undergoing a detonative phase transition could generate significant shock waves.

5.9 Dynamic Models

The thermodynamic model holds only for equilibrium systems. For all the other systems not at thermodynamic equilibrium, such as transport processes and incomplete reactions, a complete study of the dynamics of the system is necessary.

Frequently we use materials under conditions where the thermodynamic model tells us they are unstable. However, they remain quite satisfactorily in this metastable state because of the inertia to 'chemical change' (reaction to give new chemical products) or to 'phase change' (conversion to a different state without chemical reaction). We are able to prepare these materials under conditions where they are thermodynamically stable, and subsequently change the conditions rapidly. Alternatively, we may so arrange the conditions that the system never reaches equilibrium. Most of our construction materials fall into this category.

In some situations the delay in achieving an equilibrium state is due to the absence of 'nucleation' sites necessary as starting points for the new phase. An example is the tendency of water to supercool, that is, to exist in the liquid state below the normal freezing point of zero degrees Celsius[39]. In other cases the rearrangements would be impossibly slow even if nuclei existed. The number of nucleation sites where carbon dioxide molecules can gather and coalesce limits the rate of formation of the gas cavities in carbonated drinks. That the cavities increase in size as they rise is not due to the decrease in hydrostatic pressure (which is negligible). Rather, the surfaces of the cavities act as nucleation sites for further carbon dioxide to come out of solution[40].

Two other, very different, examples of metastable states also involve liquids. Some liquids become very viscous as we cool them before they crystallise. Particularly if we carry out the cooling very rapidly, the viscosity may increase without freezing taking place, resulting in a solid-like material with irregular molecular arrangement: 'glass'. The short temperature interval over which the viscosity changes from what we would describe as a 'very viscous liquid' to what we would call a 'glassy solid' is the 'glass transition temperature'. This transition temperature is not associated with an equilibrium process, so does not have the thermodynamic significance of a melting point. Over long periods of time crystals may form, the crystallisation process being called 'devitrification' in this situation. ('Vitreous' means 'glassy'.) Treacle is a viscous liquid from which we can make a glass ('toffee') by cooling. 'Fudge' is devitrified toffee: small crystals formed within a glassy structure[41].

Another example demonstrates the extent of the cohesive interactions in liquids: under suitable conditions, liquids can exist under tension or 'negative pressure'. Without nucleation sites to initiate vapour formation, it is possible for a metastable situation to develop. One can achieve these conditions in a strong pressure vessel or 'bomb' completely filled with water, heated, sealed at high temperature and allowed to cool. A flexible container would collapse as the water cools, but a pressure vessel withstands those forces and significant tensions or 'negative pressures' can be maintained in the water. In a similar way the narrow capillaries of trees and the transpiration process keep sap under tension while it rises from the roots to the tree tops[42].

Endnotes

[1]An example here is the estimation of aromatic stability (**9.13 Aromatic Model**). Further reading (particularly relating to atom clusters) Berry (1990).

[2]Particularly for chemical reaction kinetics (**12.8 Chemical Processes**).

[3]Further reading Atkins (1984).

[4]See **1.9 Starting the Jigsaw Puzzle** and **3.1 States**.

[5]These concepts are discussed in **3.2 Dynamics** and **3.7 Chaos**.

[6]'The ergodic skeleton in the cupboard of thermodynamics' as quoted by Weber (1992, p 89).

[7]See **2.6 Limits and Dimensions**.

[8]See **5.3 Absolute Zero**.

[9]See **1.12 Misleading Models**.

[10]Further reading Davies (1985, Chapter 4), Postle (1976, Chapter 9), Sparberg (1996).

[11]See **14.2 Interfacial Energy**.

[12]The calorific value (the internal energy change when burnt) of foods is usually expressed in kilojoules (1000 joules) per gram or in kilocalories (1000 calories) per gram. To give some idea of magnitudes of these units, the energy associated with each human heartbeat is approximately one joule, and one calorie is just over four (4.184) joules. Confusingly, the 'kilocalorie' is sometimes called a 'Calorie' when food calorific values are quoted. When you see 'Only two Calories!' it almost certainly means 'only two thousand calories'.

[13]See **12.9 Catalysts and Enzymes** and **14.7 Surfactants**.

[14]The investigation of such a system with which I was involved is described by Hefter *et al* (1991). See also **1.13 Phases**, **2.4 Condition Critical** and **3.1 States**.

[15]See **2.6 Limits and Dimensions**.

[16]Atkins (1984, p 42).

[17]Some systems can be considered to have **negative** temperatures **if temperature is defined in terms of the Boltzmann equation** and if populations of energy levels are manipulated to create a population inversion, with the highest energy level the most populated, for example in nuclear spin distributions: see Owers-Bradley (1993).

[18]Logarithms are discussed in **2.2 Perspective**. The value of the Boltzmann constant is $k = 1.38 \times 10^{-23} \, \text{J K}^{-1} \, \text{mol}^{-1}$.

[19]Further reading Atkins (1984).

[20]Further reading Ruelle (1991).

[21]Further reading Atkins (1984), Davies (1985, Chapter 14).

[22]See **5.8 Phase Changes**.

[23]Further reading Pippard (1966, p 99).

[24]Further reading Atkins (1984, p 21; 1990, p 105).

[25]Lovelock (1988, p 23).

[26]Greaves and Schlecht (1992).

[27]These are discussed further in **13.9 Magnetic Matter** and **12.5 Mixtures**. Differences in ordering in elementary materials give rise to allotropes (**9.6 Allotropes**) that coexist at room temperature: although there is always one form more stable, the other may persist indefinitely. See also **5.8 Phase Changes**.

[28]**11.6 Turbulence**.

[29]Kauffman (1993).

[30]See also **2.4 Condition Critical**, **3.7 Chaos**, **11.6 Turbulence** and **15.12 Life**. Further reading Atkins (1984, p 183).

[31]See also **4.2 Wave States**. Further reading Lovelock (1988, p 75), Ruelle (1991).

[32]Further reading Ouyang and Swinney (1991), Winfree (1991).

[33]Further reading Fineberg (1996), Umbanhowar *et al* (1996). See also **11.7 Granular Matter**.

[34]See **12.4 Phase Diagrams**.

[35]See next section and **11.5 Glasses**.

[36]Alloys: see **12.1 Purity**.

[37]See **13.9 Magnetic Matter** and **11.3 Superfluid Helium**. In second-order transitions the **temperature dependence** of volume and enthalpy changes at the transition temperature; they are also known as 'lambda transitions' because the plot of heat capacity against temperature is reminiscent of the shape of the Greek letter lambda.

[38]Further reading Barton *et al* (1971), Barton and Hodder (1973).

[39]See **13.3 Nucleation**. Further reading McBride (1992).

[40]Further reading Shafer and Zare (1991).

[41]See **11.5 Glasses**. Many ceramics (**13.6 Ceramics**), including those used as superconductors (**13.7 Superconductors**), are formed at high temperatures then used in a metastable state, and some of the metastable allotropes (**9.6 Allotropes**) such as diamond exist for extended times as practical materials: further reading Sleight (1991).

[42]Further reading Hayward (1971), Apfel (1972).

CHAPTER 6

INTERACTING MATTER

The Ancient Greek scientists distinguished between **forces** and matter. We often find that a Universe modelled on **energy** and matter is more convenient, although our lifelong experience with gravitation (our own weight) is 'felt' as a force. In exploring matter, we should be able to move smoothly backwards and forwards between the force model and the energy model.

6.1 Interactions

Consider the methods of visualising the attractive force between two masses[1]†. Isaac Newton originally related the magnitude of the (instantaneous) gravitational attraction directly to the masses of the objects and their separation. Special relativity proposes that nothing moves faster than the speed of light. As a result the 'classical field model' considers a field propagating the force at a finite speed. General relativity uses the model of curved spacetime (although that idea was first introduced in a mathematical model in 1913 by Gunnar Nordstrom). Here, gravitationally attracted objects follow the shortest path in four-dimensional spacetime, which appears to us as a curved path in our three space dimensions. (It is not appropriate to say that the planetary bodies follow the curved paths they do because of gravitational interactions. Only in the newtonian model is gravity a force like other forces. The small deviations from classical or Newtonian

† Endnotes for this chapter can be found on page 142.

predictions in the orbits of the planets are consistent with general relativity.) Models to help us visualise the situation include

- a flexible canvas in four dimensions on which our activities are painted, and

- an aerial view of a landscape obtained by an aircraft pilot, with the extent of discernible detail varying with height.

The relativity model treats spacetime as continuous (and separable into 'space' and 'time' only on a macroscopic scale)[2].

Because energy is equivalent to mass, it is reasonable that energy is also subject to gravitational effects. The Apollo mission confirmed this in the Earth–Moon–Sun system by showing that the gravitational attraction of the Sun acts on gravitational energies between masses on Earth[3]. We can also see gravitation affecting a starbeam passing close to the Sun on its way to the Earth. The Sun's gravity deflects the light and we can detect the shift in the apparent position of the star (seen during a solar eclipse).

Consider the 'fundamental interactions':

- gravitation and electromagnetism, which are familiar,

- the 'weak interaction', which is important in nuclear decay, and

- the 'strong interaction', which binds atomic nuclei.

As explained later[4], in the force-carrying particle model for all these fundamental interactions the range is inversely proportional to the mass. The effect is short range for the weak interaction but infinite for the massless particles such as electromagnetism and gravitation. In these latter cases the interactions fall off inversely as the square of distance (an inverse-square relationship).

As well as range, the nature (attractive or repulsive) of the interaction is of fundamental importance. It is attractive only in the case of even-integer spin fields such as gravitation but can be either attractive or repulsive for odd-integer spin fields like electromagnetism[5].

In the model for molecular interactions, we can most easily visualise the net effect of the fundamental interactions as a balance between attractive (or cohesive) and repulsive interactions. Both are relatively short range, but the magnitude of the repulsive interaction increases rapidly as atoms approach each other closely. In the other direction, as the interatomic distance increases, the attractive interaction drops off more gradually, to become negligible at several atomic diameters (at the order of a nanometre distance). This force model and the energy model are alternative ways of describing the same phenomena[6].

6.2 Probing Matter

Chemists and physicists can investigate matter by firing particles with known properties into an object and measuring what comes out as a result of the interactions or collisions. If it is more convenient, we can treat the probe 'particles' as 'waves'. It is often helpful to consider energy propagating as waves and being emitted from or absorbed by matter as particles[7]. There are several groups of experimental techniques available for studying various materials:

- microscopy, generating a visible image of a smaller object

- spectroscopy, measuring energies absorbed or emitted

- diffraction, modelled as the interaction of waves with a periodic 'lattice' generating an image in 'reciprocal space'[8]

- 'refraction', caused by the change in velocity of waves due to their interaction with matter of any kind.

Some of these effects are apparent to us without any special equipment, particularly our direct perception of the 'colour' of objects (a form of what scientists call 'spectroscopy' when they measure this property with instruments). Our perception of the colour of a substance can be explained in terms of the absorption, transmission and reflection of certain portions of the visible spectrum of light. Another characterisitc property is its 'refractive index'. This is the retarding effect it has on light, and is expressed as the ratio of the speed of light in a vacuum, to the speed of light in the substance. In some materials, the refractive index is different in various directions. This imbalance rotates plane-polarised light and results in coloured 'birefringent' (doubly refracting) patterns being visible. Coloured interference effects are also common.

Visible light scatters at the surface of an object if the structure there has the dimensions of the wavelength of light. As glass is polished to make mirrors and lenses, for example, the scratches on the surface become progressively finer until they become too small to interact with the light. The glass is then transparent to those wavelengths: it becomes 'clear'.

The size of the particles we can 'see' depends on the wavelength of the particle we use as a probe[9]. Because the wavelength of visible light is much larger than the size of an atom, we cannot see an individual atom by 'shining' light on it. The wavelength of a particle depends on its energy, so the higher the energy of a particle, the smaller the object it can probe. Before the twentieth century, only low particle energies were available (energy generated by chemical reactions such as burning carbon compounds), and the atom was the smallest particle that we could see. Physicists can now accelerate electrons to energies billions of times greater, and they 'see' far smaller particles[10].

Postle[11] has used the following analogy. I am trying to make a detailed model of the outline of one of my hands, placed with fingers spread palm-down on a table, by sprinkling it with particles of different sizes. I find that anything larger than grains of rice gives a very poor model. Sand is better, but if I use a fine material like flour I can make a very detailed image.

There is another way of 'seeing' objects far smaller than the wavelength of light using an ordinary optical microscope: the method of 'decoration'. We sometimes make the position of a wire or rope more visible for safety reasons by hanging streamers from it. In a similar way, we can 'see' sub-microscopic phenomena such as surface defects. We arrange for them to act as nuclei for the deposition of larger and more visible particles or for the dissolution of visible etch pits on the faces of crystalline solids.

The interaction of molecular particles (atoms or molecules) with 'light' (visible or near-visible electromagnetic energy) is an important part of chemistry, from both theoretical and practical points of view. Here it is convenient to use the photon model, and say that a molecular particle in a low energy state (the ground state) absorbs a photon, with the result that the particle is in an 'excited state'. There is also the reverse process, which is 'fluorescence': a molecule in an excited state emits a photon, so the molecule is in the ground state. This field of science or modelling is 'molecular quantum electrodynamics'. With it physicists discuss not only the interaction of light with molecules, but also the interaction of molecules with each other: 'intermolecular forces'[12].

At another level, with other models, we say that electrons probe the 'inside' of protons and neutrons and 'see' quarks. More precisely, the large angles through which the electrons are scattered reveal charged point-like particles within protons and neutrons. This parallels earlier experiments where the nuclei within gold atoms in gold foil scattered alpha particles through large angles. Because the wavelength is inversely proportional to momentum, and the quarks are very 'small', we can use only very high-energy electrons for this purpose. However, high-energy probes are likely to have collisions causing irreversible changes[13].

The effects resulting from interaction of radiation with matter can be subatomic, atomic, molecular or macroscopic. Coloured oil films on water provide a common example of interference. Here there is interaction of light waves with an oil film about a light wavelength thick. In this example it is convenient to use the wave model. For certain wavelengths of light, the crests of the waves reflected from one side of the layer coincide with the troughs reflected from the other, so the waves cancel. These wavelengths are absent from the reflected light and we see the complementary colour. We see colours in soap bubbles for the same reason[14].

If light is directed at two convex lenses placed in contact under load, circular fringes appear in the thin air space around the glass-to-glass contact. In the centre there is a very small black spot where there is such intimate contact between the solids that they have effectively fused together locally. We can observe this more easily with polymer glasses than with silicate glasses, because polymers are both more flexible and less likely to adsorb[15] contaminants such as water.

Sometimes we do not need to see all the individual atoms to obtain information on a structure or process. The way in which the protein haemoglobin traps and releases a molecule of carbon monoxide was investigated[16] by illumination with simultaneous laser pulses to give an interference pattern. In the 'light' areas the carbon monoxide molecules escaped, and the associated change in shape of the protein generated a sound wave detected by a longer-wavelength laser pulse.

Spectroscopy measures the changes in energy of radiation entering or leaving a material. Chemists and physicists have been very creative in developing an armoury of spectroscopic techniques spanning a wide range of energies and the whole electromagnetic spectrum. Distance is little barrier. It is possible to investigate the composition of the stars and of interstellar space in this way, by studying their own radiation. The basis of spectroscopy is that the quantised energy gained or lost by a particle appears as a shift in the energy (and therefore the frequency) of the associated radiation.

Waves passing through a periodic array of atoms or molecules (a lattice structure) with repeat distances comparable to the wavelength produce a pattern of high-intensity points. This is useful, because the particular pattern depends on the nature of the lattice. Researchers have applied this experimental technique most extensively in x-ray diffraction (because the wavelengths of x-rays are comparable to spacings in crystals) and in electron diffraction. The phenomenon described by this model is general, however. In terms of the wave model, each layer of molecular particles acts as a partial mirror, reflecting the incident waves weakly. If the reflected waves from each plane are in phase (which occurs at particular orientations of the planes to the incident radiation), there is reinforcement. An observable pattern characteristic of the symmetry of the crystal results when the radiation strikes a detector. The effect of this is to generate an image in 'reciprocal space'[17].

If we direct high-energy particles along a crystal plane of symmetry of a solid (rather than at an angle to it) the particles tend to be 'channelled' instead of interacting randomly with atoms. They 'see' long strings of atoms and interact with them as a unit. The result is that the particles penetrate more deeply and emit radiation in a different way because of the different interaction. We must distinguish between 'light' and 'heavy' particles, the former (like electrons and positrons) emitting far more energy as radiation than heavy particles (like protons). A classical (rather than quantum mechanical) model for the interaction is of marbles running down an inclined sheet of corrugated iron. They tend to be constrained to the troughs if they are initially reasonably parallel to them (that is, unless they have too much transverse energy). This model takes us a fair way towards understanding channelling, before we look at the quantum mechanical properties. Positive and negative particles interact very differently with the lattice of positive nuclei. The troughs in our corrugated iron model differ in the two situations. For positively charged particles that move between the rows of positive nuclei in the crystal the troughs have flattish bottoms and steep sides. The positive nuclei are positioned along the ridges. For negative particles, on the other hand, the positive nuclei provide deep, steep troughs for interaction.

The quantum mechanical model tells us the charged particles are not localised, and they can have only certain energy states. For heavy particles, the quantum states are closely spaced and their behaviour approximates to classical. Even for electrons, predictions based on classical mechanics are reasonable in regions close to the nuclei. At very high particle energies, effects associated with relativity become important. One of these is that particles have higher relativistic mass, resulting in more classical behaviour because the quantum states are closer. There is also a relativistic increase in the frequency of emitted radiation. Applications of particle channelling include sources of tunable x-rays and gamma radiation[18].

Electronic 'insulators' and 'semiconductors' diffract the outer or bonding electrons[19] (modelled here as waves with wavelengths comparable to atomic spacings) into closed bands and so prevent the electronic conduction that occurs in metals. A similar phenomenon occurs on the macroscopic scale. We can construct a dielectric material with a lattice of holes a few millimetres apart which interact with microwave radiation of that wavelength in a similar way. The system acts as a microwave insulator for the 'forbidden' range of microwave wavelengths. There is even evidence that it is possible to construct a lattice of holes on a much smaller scale to interact selectively with visible radiation. This would be a 'light insulator' or 'photonic insulator' that would 'trap' a photon. The nearest natural equivalent may be opals, their brilliant colours resulting from diffraction effects. In principle it would be possible to develop a semiconductor laser device, with electron bands determined by atom spacings controlling electronic properties and a lattice structure etched at much wider spacings to control photonic properties. There would be sufficient disparity of scale between the atom-scale electronic lattice and the etched photonic lattice to prevent one from influencing the other[20].

The scattering of radiation from small inhomogeneities provides information on the sizes of particles and the internal natures of structures. 'Rayleigh scattering' of visible light occurs if the structures are smaller than the light wavelengths. The scattering of blue light (wavelength around four hundred nanometres) is ten times greater than that of red light (wavelength of the order of seven hundred nanometres). (To put these distances in context, remember that most **molecular** particle diameters are less than one nanometre.) The scattering by particles in the atmosphere is the reason for the Sun appearing to be red or yellow, particularly when rising or setting, and for the sky during the day being blue.

In some situations it is better to use larger particles as projectiles. Surface physicists accelerate molecules of the fullerene C_{60} to high energies and use them used to dislodge biomolecules adsorbed on surfaces for subsequent analysis by mass spectrometry[21].

6.3 Transformation

The Fourier transform is a mathematical procedure with various applications. One is the 'unscrambling' of x-ray diffraction

patterns obtained in 'reciprocal space' from crystalline solids.
They are transformed into information on the way the atoms or
molecules are arranged in the crystals in 'real' space. Once a
Fourier transform is used on diffraction results, one can choose to
avoid even thinking about reciprocal space.

It is informative to approach Fourier transformation by means of
the 'model' used by Jean-Baptiste-Joseph Fourier himself in the
early nineteenth century. In this case we are using a real life
object to illustrate a mathematical concept, rather than the
reverse. One-half of an iron ring was heated to red-heat in a fire,
then the ring was buried in insulating sand and the temperature
measured around its outer circumference. As heat was conducted
from the hot to the cool region, the temperature varied around
the ring. When he measured the temperature and plotted it on a
graph he saw an 'S'-shaped or sinusoidal wave. Over a period of
time as the iron cooled the variation became less marked, then
eventually flattened to uniformity. Fourier saw that he could put
together the initial 'step' or 'sawtooth' function in temperature
around the ring from a number of sinusoidal waves. Each
sinusoidal wave varied from minimum to maximum and back
again an integral number of times (one, two, three, etc) around
the ring. The 'fundamental harmonic' wave that varied only once
around the ring decayed less rapidly than the higher 'harmonics'.
In principle an infinite number of harmonic waves is necessary to
duplicate the original step function, but in practice a good
approximation does not require many harmonics. The Fourier
transform of the 'step' is the mathematical function that describes
the amplitude (temperature in this case) and phase (position of
the stationary wave around the ring) for the set of harmonic
waves[22].

Our ears do the same kind of thing in reacting to sound waves,
although we are unaware of it. They convert the time-dependent
pressure fluctuations (pressure waves or sound waves) in the air
into what we hear: a series of distinct frequencies each with its
own intensity or volume.

When x-rays interact with a molecular crystal a 'diffraction
pattern' is 'created' by the system itself. The periodic electron

density within the crystal structure transforms the incident x-ray beam into an array of scattered beams that chemists and physicists monitor by generating spots on a photographic film or recording instrumental readings. What they then do is work backwards from the diffraction pattern so that they can deduce the original molecular structure. The amplitude of the wave is accessible from the intensities of the spots, but indirect methods are necessary to obtain the phase information. The result is a three-dimensional map of electron density in the molecule that provides information on atomic positions and therefore the molecular structure.

We can illustrate the diffraction process experimentally and make the subsequent transformation by using an 'optical transform'. Shining visible light through an ordered array of dots on a projector slide is analogous to directing x-rays through a crystal lattice, and creates a similar image[23].

6.4 Changing Matter

The interaction of radiation with matter may be reversible, or it may cause permanent change.

At the high-energy end of the scale physicists use particle accelerators to search for smaller and smaller subatomic particles of fundamental matter, but here we consider the effect of more modestly energetic interactions of radiation with molecular matter.

Chemical changes caused by light are 'photochemical' processes or 'photolysis'. If ionising radiation is the energy source (resulting in loss of electrons from the molecules and the formation of radicals and charged ions) we call the process 'radiolysis'. Because water is present in such large amounts, the effect of ionising radiation on the living cell is mainly the result of chemical processes following the formation of hydrogen atoms, hydroxyl radicals and hydrated electrons. (It is worth noting that the highly reactive molecular oxygen causes similar effects, although oxygen is also essential to the survival of the cell[24].)

There are also more subtle effects. 'Photochromic' materials (now used as lenses in some sunglasses) change colour when light passes through them, so changing the nature and amount of light absorbed. In 'photorefractive' materials the light alters the refractive index of the materials (to an extent that depends on the light intensity). Both of these 'nonlinear' optical effects, photochromism and photorefractivity, are in contrast to the behaviour of 'linear optical media' like ordinary lenses and prisms through which light passes without changing the material in any noticeable way. In some cases these effects last until we deliberately reverse them, providing another method for the storage of information. One model for the photorefractive effect is that of an ant moving a sandhill one grain at a time. A low-intensity beam of light gradually builds up an electric field by shifting charges associated with crystal defects one electron at a time, leaving a positive charge behind. This continues until the field is strong enough to distort the crystal lattice very slightly (about 0.01%) but sufficient to change the refractive index. The charges trapped at defects can move only in the presence of light, so if two laser beams interfere to form a pattern, charges migrate from light regions and are trapped in dark regions. This results in the refractive index changing periodically, and can generate a refractive index grating or hologram within the crystal. One of the effects is 'beam fanning', a spreading and curving of a laser beam into a broad fan of light.

Practical applications of these effects include

- protection of light-sensitive detectors from laser beams,

- communication of information through a distorting atmosphere, and

- 'novelty filters'.

'Novelty' here has its original meaning of 'new'. These devices highlight moving images and ignore stationary backgrounds in applications such as the detection of moving aircraft against a complex background or of living microorganisms amongst inanimate debris under a microscope[25].

Other applications of laser beams are for 'cooling' atoms (reducing their energies), as 'optical traps' and as 'optical tweezers' for molecules or particles up to the size of a micron (a thousand nanometres). The intense electromagnetic field in a tightly focused laser beam polarises particles like bacteria, or components within a living cell, or even strands of DNA, and allows us to manipulate them by moving the beam. It appears that we can do this without damage to the particles as long as they are transparent to the light being used[26].

6.5 Failure

When forces act on bulk materials the nature of their responses is important when engineers are deciding how to use the materials. A material may flow irreversibly, or it may fracture irreversibly. (Or, if the force is not too great, it may bounce back to its original configuration.) Both failure mechanisms, flowing and fracturing, exist in all materials to different extents. If a particular material yields before it cracks we describe it as 'ductile' or 'plastic' while if it cracks first we say it is 'brittle'[27]. We can make apparently 'brittle' materials fail in a flow manner by using the appropriate conditions, such as during the compression of powders into tablets.

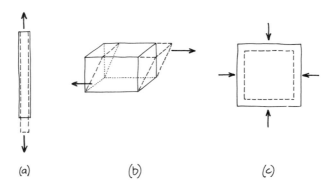

(a)   (b)   (c)

Figure 6.1 The three elastic moduli. (a) Young's modulus, relating stretching to equal and opposite forces at either end of a rod. (b) Shear modulus, a torque produced by equal and opposite forces applied to opposite faces of an object. (c) Bulk modulus, the volume change caused by a uniform force applied over the whole object (a change in pressure). This is illustrated in two dimensions for simplicity.

If the deformation process is reversible so that the material recovers its size and shape when we remove the force, we say the material is 'elastic' under the particular conditions. If the elastic displacement is proportional to the deforming force, we say that the material obeys 'Hooke's law'. Hooke's law is an accurate description over a narrower stress range than the range of elastic behaviour. In other words, materials can be elastic without the deformation being exactly proportional to the stress. Both elasticity and Hooke's law behaviour can be related back to the energy well model of the chemical bond[28]. (So far I have assumed that the elastic properties are independent of how rapidly we apply the stress. This is not so, and materials, solids as well as liquids, show 'viscoelastic' properties under certain conditions)[29].

We measure elastic properties by three quantitative measures of behaviour: the 'elastic moduli' (figure 6.1). Up to the 'elastic limit' of the material where permanent change begins to occur,

- 'Young's modulus' describes the tensile or stretching properties

- 'shear modulus' or 'rigidity' describes the reaction to tangential forces generated in twisting situations

- 'bulk modulus' measures the behaviour under compression.

We can describe all types of deformation as various combinations of length change, shear and volume change. Crystalline solids have measurable values of all three moduli. Fluids usually possess only a bulk modulus, but in certain conditions can withstand tension[30].

We 'normalise' the effect of loads on a material and the effects they have by expressing them per unit length or unit area or unit volume. The word 'stress' has a more precise meaning than in general use. For example, the tensile stress is the applied load per unit area and the 'strain' is the extension per unit length. When we plot a graph of the measured strain as a function of stress we apply to a sample, we get a curve like figure 6.2. There is a linear, 'elastic' region up to the point E, the 'elastic limit' of the material. Behaviour in the region described by the line OE is reversible,

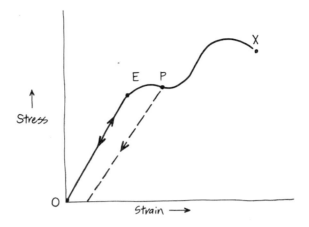

Figure 6.2 A stress–strain curve for a tensile test (measuring the stretching as a result of the force applied at opposite ends of a rod). E is the elastic limit, and Young's modulus is given by the slope of the line OE. Plastic flow occurs in the region EX, and fracture occurs at X. In some materials the elastic region OE is very short, and in some the plastic region EX is very short.

and the slope of this line defines the Young's modulus of the material. If we increase the stress beyond this, say to point P, plastic flow or deformation occurs, and when we release the stress, permanent change has occurred. Further applied stress results in the sample breaking at point X. This is the 'breaking stress' or 'tensile strength' or 'ultimate strength'. Shear and compression tests show similar behaviour.

An 'elastic material' has properties that are 'ideal' or 'limiting'[31], a model material with associated graphical or numerical models yielding the three elastic quantities. When they combine these values with the 'ultimate' or 'breaking' characteristics, engineers have a very useful system to describe real materials. This is not the only way of modelling the response of materials to stress, but it is simple and adequate for most practical purposes. That is all we ask of a model.

In descriptive terms we characterise a material by the extent of its tensile strength (or the converse property, weakness) under tension and its stiffness (or flexibility) under shear. A metal like steel is both strong and stiff. Polymers tend to be strong and flexible. A jelly or 'gel' is flexible and weak. Also important is the 'toughness', which is the resistance to failure by cracking. The bulk modulus or compressibility of the condensed states (liquids and solids) is far greater than that of gases, which are very compressible. Some of these materials behave so differently from elastic materials that we call them 'non-newtonian'[32].

Bulk crystalline solids are not homogeneous and uniform, but are made up of crystalline grains. Between crystals of different orientation there are grain boundaries, as illustrated in figure 6.3. Also, there are often assemblies of microcrystalline materials of

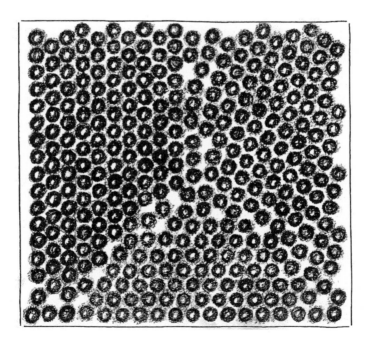

Figure 6.3 Grain boundaries in a polycrystalline material. (After Geselbracht *et al* (1994).)

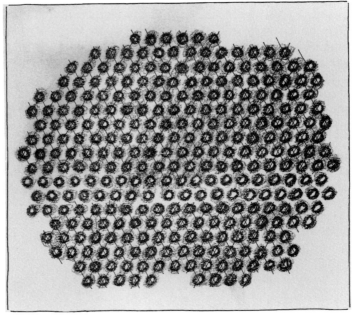

Figure 6.4 One plane of a crystal dislocation. The atoms joined by lines form regular arrays, and the other atoms form the dislocation. (After Geselbracht *et al* (1994).)

different structure or composition, separated by phase boundaries. The crystals themselves contain dislocations—imperfections that can move around (allowing plastic flow) or entangle (leading to hardening) (figure 6.4).

We can model a two-dimensional imperfect crystalline array with a bubble raft: generating uniform bubbles on the surface of an aqueous detergent solution. (We can make a simple version of this in the kitchen sink with a tap dripping regularly into clean water containing a little detergent.) 'Grain boundaries' arise naturally between 'grains', arrays of bubbles with regular hexagonal (six-fold) symmetry. We can move 'defects' (for example, vacancies created by popping one bubble) around within a uniform grain by deforming the array with a slider[33].

Thin pieces of material and particularly single-crystal 'whiskers' are often tougher than bulk samples because of the absence of boundaries between crystal grains and of surface dislocations[34]. Alternatively, some materials which are not inherently tough can be made tougher by 'pinning' the dislocations. We do this by

- introducing impurity atoms to impede the movement of dislocations (like a small proportion of carbon atoms amongst the iron atoms in steel), or

- entangling the dislocations (for some metals by cold-drawing them through a die).

Techniques such as electron microscopy provide information on how dislocations and interfaces determine material failure.

Endnotes

[1]Further reading Goldman *et al* (1988). See **3.3 Newton's Model** and **4.6 Quantum Leap**.

[2]See **2.6 Limits and Dimensions**. Further reading Davies (1980, Chapter 5).

[3]Further reading Hawking (1988), Goldman *et al* (1988).

[4]See **7.8 Colourful Quarks**.

[5]See **4.5 Playing Cards and Four-Strokes**.

[6]Discussion of interactions between molecules is continued in **10.1 Molecular Interactions**. See also **8.1 Atom Models**.

[7]Further reading Postle (1976, Chapter 6).

[8]Discussed below. See also **6.3 Transformation**.

[9]See **3.6 Light** for the description of coloured light in terms of a spectrum or spectral distribution of wavelengths. In describing the interaction process, we are combining the terms **wavelength** and **particle**, as introduced in **4.1 Ultraviolet Catastrophe** and **4.2 Wave States** (even though this approach may be initially unfamiliar) so that we can use the most appropriate scientific model for the topic being discussed.

[10]The electron volt (eV), although not compatible with SI units, is a very easy energy unit to appreciate when discussing atoms and electrons, being the energy one electron gains when it is accelerated by a potential difference of one volt. Chemical reactions produce particle energies of a few electron volts; Rutherford used alpha particles with energies of millions of electron volts (MeV); and we now use electromagnetic fields providing thousands of millions of electron volts. The mass of an electron in energy units (see **4.6 Quantum Leap**) is 0.5 MeV and the mass of a proton is over 900 MeV.

[11]Postle (1976, Chapter 6).

[12]Further reading Craig (1992). Molecular interactions are discussed in Chapter 10.

[13]Further reading Ryder (1992), Farmelo (1992), Postle (1976, Chapter 6).

[14]'Thermochromic' liquid crystals (**15.8 Liquid Crystals**) display varying colours with changing temperature as molecular dimensions of the order of visible light wavelength change.

[15]See **14.4 Adsorption**.

[16]Hecht (1991a).

[17] **6.3 Transformation**. Further reading McPherson (1989), Moffat (1992), (1991), Elliott (1991). See also **2.8 Symmetry**.

ing Sorensen and Uggerhoj (1989). See also **4.6 Quantum Leap**.

d Models.

[20]Further reading Pendry (1991). See also **2.2 Perspective** and **9.5 Classifying Matter**.

[21]Further reading Aldous (1992).

[22]Further reading Bracewell (1989), Moffat (1992). See also **6.2 Probing Matter** and **13.2 Crystals**.

[23]Further reading Lisensky *et al* (1991).

[24]Further reading Lovelock (1988, Chapter 7).

[25]Further reading Pepper *et al* (1990). See also **4.2 Wave States**.

[26]Further reading Ashkin (1972), Chu (1992). See also **8.8 Cooling Atoms**.

[27]Further reading Gordon (1968).

[28]See **5.1 Energy Well Model**. Further reading Gordon (1968, Chapter 2).

[29]See **14.13 Aggregates**. The processes occurring during flow are described in **11.4 Viscous Flow**.

[30]See **5.9 Dynamic Models**. In many situations in so-called soft materials (**15.1 Soft Matter**) there is a combination of viscous and elastic processes ('viscoelasticity') and the rate at which the applied force changes is particularly important, as discussed in **15.4 Flowing Polymers**.

[31]See **2.6 Limits and Dimensions**.

[32]See **11.4 Viscous Flow**.

[33]Geselbracht *et al* (1994).

[34]Selinger (1989).

CHAPTER 7

FUNDAMENTAL MATTER

Chemists and physicists have devised numerous models that account for enough of the observed properties of materials to be acceptable approximations to reality. The linear or sequential form necessary for writing an account of materials causes problems in presenting these interlocking models[1]†, but I shall start with the smallest 'particles' and gradually move up to macroscopic materials.

7.1 Beyond the Atom

Our perception of everyday materials is that at high temperatures substances are gaseous, with the molecular particles (atoms or molecules) moving around at high speed, colliding randomly, and filling any containers they occupy. As the temperature falls (as the internal energy decreases), there is a range of conditions in which the intermolecular cohesive interactions balance the repulsive interactions so the material condenses into a liquid. At a still lower temperature it freezes into a crystalline solid in which the position of every particle is essentially fixed. Alternatively, consider the reverse process. Ice cubes (solid water) left in a warm room absorb energy, melt to form liquid water, which absorbs further energy and eventually evaporates as water vapour[2]. If one wishes to visualise it in this way, the states of matter result from competition

---

† Endnotes for this chapter can be found on page 168.

between thermal energy and the energy of molecular interactions. Liquids and solids are condensed matter; liquids and gases are fluid matter. Sometimes it is helpful to use pictorial analogies[3], such as

- the ordering of a military unit for solids,

- the dynamic close-packed disorder of a reunion party for liquids,

- the lower-density disorder of a soccer game for gases.

The picture just described greatly over-simplifies the situation for all materials, and some materials do not conform even to this broad outline. For example, helium, composed of identical and chemically inactive atoms, remains a liquid even at the lowest temperatures and its properties are fundamentally different from those one might expect[4]. It is therefore not surprising that materials composed of more complex particles than single atoms also show unexpected behaviours. To make progress in appreciating the properties of materials, we now explore subatomic models, and move away from the image of an atom as a small hard sphere.

In modelling the fundamental basis of materials it is helpful to make use of an idea already introduced. Let us assume various fundamental fields are always in existence, in some situations interacting as if they were particles. We focus our attention on the fields and interactions rather than on the particles, on the behaviour rather than on the composition[5]. Interactions appear to be of only four kinds, with an atom being the result of:

- an 'electromagnetic interaction' between fields associated with the nucleus and the electrons,

- a nucleus arising from the strong interaction between neutron and proton fields,

- the fields of all particles interacting through a gravitational interaction.

- There is also a weak interaction.

Even in accepting this model, however, we retain the concept of 'particle' in those situations where it is appropriate (which for macroscopic objects in everyday life is most of the time). So we continue talking about particles as solid, discrete objects. At the same time we should remember that sometimes particles may be transient and better described as interactions between fields.

We can distinguish three aspects of materials.

- Particles or 'building blocks'.

- Structures into which the particles are assembled (permanent or temporary on a human time scale).

- Forces, fields or other convenient interaction models holding the particles in the structure.

At one extreme of very small particles, we may model the 'building blocks' as 'quarks' held together by 'gluons' into a 'structure', the nucleus. Alternatively, at the next level up in size, we can build 'atoms' from 'electrons', 'protons' and 'neutrons'. Molecules are assembled from the 'building blocks' of atoms by means of 'chemical bonds'. Moving up to the macroscopic scale 'structures' can be arrays of molecular particles (molecules or atoms or ions) exhibiting 'physical' interactions, which in turn may be 'domains' or 'crystallites' packed together into larger 'structures'. This technique of building up structures from smaller building blocks is just one of the ways of classifying the enormous array of information available to us. It has been successful throughout the history of science. Some people have concentrated on the properties of individual particles: molecules atoms or subatomic particles (in a 'reductionist' or 'bottom-up' manner). Others have dealt with bulk materials (a 'holistic' or 'top-down' point of view). These approaches are complementary rather than competing. To emphasise this fact it is worth noting as an intermediate situation that microclusters occur, with properties different from those of bulk matter but more complex than those of individual atoms or molecules[6].

7.2 Bubble Chambers

The vapour trail left by a high-flying aircraft provides a relatively long-lasting impression of its passage, and we can make a

permanent record by photographing the vapour trail. The 'bubble chamber' does the same thing for the very much smaller, subatomic, relativistic particles.

A bubble chamber is a tank of fluid in a thermodynamically metastable[7] condition such that high-energy charged particles nucleate vaporisation as they pass. Physicists maintain a superheated neon–hydrogen mixture at low temperature (about 30 kelvin or minus 240 degrees Celsius) under pressure, then expand it suddenly just before use. The particles under investigation cross the liquid phase in a few nanoseconds. The bubbles nucleated take a million times longer (about ten milliseconds) to expand to a size large enough to photograph. (This is done from more than one angle to obtain three-dimensional tracks.) The bubble chamber is usually within a strong magnetic field, and the resulting curvature of each track provides information on the momentum and charge of the particle making it. Although faster and more efficient particle detectors have superseded them, bubble chambers provide a more intimate record of subatomic phenomena. They enable us to get a better feel for subatomic particles to facilitate our modelling of matter[8].

7.3 Charge Integrity

When John Dalton argued for the atomic nature of matter[9], he was using a model in which mass was quantised, even though electricity was still modelled as a continuous fluid. From experiments on the chemical effects of electricity ('electrolysis') George Johnstone Stoney in 1874 proposed the model of the charge carrier as a particle. He gave it the name 'electron' from the Greek word for amber (because amber becomes electrified when rubbed with a material such as wool).

For some purposes it is still convenient for us to refer to electricity as a continuous fluid and our terminology reflects this: electric 'current' 'flows' in a 'circuit'. In other situations the electron as the quantum of electric charge is a far more useful model, with electricity being regarded as a stream of individual electrons, each carrying unit negative charge[10].

At first sight it might appear that the amount of electric charge in a conductor exists in increments of one electron charge, but this is not always so. Experiments at temperatures near absolute zero, on small electrical conductors (ten to a hundred nanometres), and extremely small currents, show that electric charge flows continuously. The charge at any time on a conductor and an insulator can correspond to a fraction of an electron rather than having to be in whole electron units.

(This is an example of the nature of bulk matter, and can be 'understood' or appreciated quite easily. Charge does not accumulate in a conductor, and the amount of charge transferred past any point depends on the sum of the shifts of all the electrons with respect to the lattice of all the atoms. Therefore it need not be a whole number of electrons.)

In contrast to what happens in a **conductor**, when electrons cross a thin (typically one nanometre) **insulating layer** they do so in a discrete way, in integral electron charge amounts, the process known as tunnelling[11]. An analogy used[12] is that charge flows continuously through a conductor like water through a pipe, but it passes across an insulating gas as discrete electrons, like drops of water from a dripping tap.

We should now be quite comfortable with these dual models of the electron as a particle and flows of electrons or electricity as a continuous current[13].

In the combination of fundamental particles to make an atom, the total mass is smaller than the sum of the masses by an amount equivalent to the binding energy. (Einstein defined this in his famous statement of the equivalence of energy and matter.) But as far as we can tell, the total electrical charge is exactly the sum of its components. (Scientists have looked at this carefully, because an extremely small difference in the magnitude of the electron and proton charges would provide a model that ultimately would account for the expansion of the Universe.) There is still interest in determining whether the charges differ between the proton and the 'antiproton' (the particle with the same properties as the proton, but with opposite electrical charge). From all this work there is no evidence suggesting that 'charge quantisation' is anything but exact[14].

7.4 Electron in a Box

Looked at superficially, the best model for an electron as a particle appears to be a very simple object. It has particular values of mass (less than a thousandth the mass of the lightest atom, hydrogen) and of electric charge, angular momentum and magnetic moment (the twisting response to an applied external magnetic field). But as a quantum particle, the electron appears very different from an ordinary macroscopic particle, its behaviour being described as a 'wave function'[15] in the quantum mechanical model. This is analogous to the behaviour of light. On the large scale, the photon or particle model is appropriate, with light travelling in straight lines. When the photons move through small spaces (small compared with their wavelength) interference and other wave-like properties appear. On the macroscopic scale, we can model electrons travelling as particles, on definite paths.

However, on a small scale, such as in an atom, the electron exhibits wave-like behaviour, with quantised energy levels. Our wave model describes this is as happening when we confine the electron in a region about the size of its characteristic wavelength. Like a wave or vibration of a violin string, the wavelength of an electron must match the space in which it is confined. To be visible by optical absorption (to be 'coloured') the dimensions have to be of the order of ten nanometres[16]. The larger the momentum of a particle, the smaller its wavelength, so we can use very high-energy electrons from particle accelerators as probes small enough to 'see' quarks within protons[17].

Probability is also valuable in thinking about electrons, so yet another model visualises the probability of finding an electron associated with a nucleus in an atom. We view the electron not as a charged particle but as a cloud of charge, the electron cloud. The 'charge density' (a measure of the electron probability) varies with distance from the centre of the atom, and we can specify the probability of finding the electron anywhere (from inside the nucleus to infinity). We note that according to the uncertainty principle we cannot say that an electron of definite energy is at a particular location. Specifying the charge density at a certain point is equivalent to knowing the probability of finding the electron there, information that arises from the wave mechanical model of the atom.

We should feel free to use whichever model of an electron is most useful for the particular circumstances (particle, or charge cloud, or wave, or probability). We must, however, remember the limitations on each model.

It is possible to confine a single electron in a 'trap' by making use of its interactions with electrical and magnetic fields. For some purposes it is convenient to treat the combination of trap and confined electron or electrons as an 'artificial atom'. This is an atom with one or more electrons and a massive 'nucleus' comprising not only the trap but also the whole of the Earth itself. An atom like hydrogen has a small, massive nucleus surrounded by an electron. This artificial atom has a large massive nucleus (the Earth) with an electron localised at a particular place on its surface. It has been named 'geonium', an Earth atom. The geonium nucleus binds its electron, as in an ordinary atom, and the properties of the electron are quantised, with a spectrum of discrete energies we can study[18]. We must practise visualising and developing models for ourselves in this way so that we can achieve the maximum flexibility of viewpoint, a process also described as 'lateral thinking'[19].

7.5 Tunnelling

Previously I introduced the idea of the tunnelling of quantum particles, referring to a tortoise-shaped wave packet and Alice stepping through the looking glass[20]. In searching for ways to visualise the properties of electrons, we can also extend our water storage lake or energy well model.

Envisage a potential energy surface accommodating electrons, just as a lake bed contains lake water. As classical objects, electrons would have sharp boundaries, as the water seems to have clear boundaries with the air and with the land. But what happens in an actual lake? There is a continual exchange of water between the lake and the air. This is most obvious if there is a morning mist on the surface of the lake, or if a storm generates waves that throw spray in the air and break over the shore. In addition, not only does rainwater flow into the lake, but water also seeps or 'tunnels' into the groundwater channels of the surrounding land. This is analogous to the wave description of electrons. Not only do they have no sharp boundaries, but also there is an element of uncertainty about their positions and they can appear outside their classical limits. We can model electron tunnelling under the influence of an applied electrical potential as the flow of groundwater between two lakes, driven by the gravitational potential of different heights.

A very similar model is that of a skateboard rider in a saucer-shaped arena[21]. In the classical model, the height the skateboarder reaches depends on the total energy (kinetic energy being exchanged for potential energy during the ride up towards the outer rim of the arena). The total energy of the rider cannot exceed that on entering the arena, so it is not possible for the skateboarder to leave. In contrast, in the quantum model, by tunnelling the rider may occasionally move past that limit and appear above the arena. However, the uncertainty principle limits the time the skateboarder can spend there. The greater the energy barrier that is being tunnelled through, the less effective is the process, and the shorter the proportion of time the skateboarder can spend out of the arena.

Another interpretation we can use is that the uncertainty effect allows small particles like electrons to 'borrow' energy for short times, enabling them to carry out these manoeuvres[22]. Although this is only a model, the electron tunnelling **effect** is real. It is of practical importance that as integrated circuits become ever smaller, component dimensions approach a size comparable to the electron wavelength. This results in electron leakage due to tunnelling adversely affecting the performance of electronic devices[23].

At low temperatures (used to avoid thermal fluctuations) a small current flowing through a tunnel junction consisting of two conductors separated by a nanometre-scale insulator generates an oscillating electrical potential. We call this 'single-electron tunnelling' or 'SET oscillation'. It is the result of a continuous flow of current through the conductors combined with a discrete transfer of charge in one-electron tunnelling steps across the thin insulator.

The charge on the surface of the conductors can be any fraction of an electron charge for the reasons explained above. If the charge on the interfaces can be reduced (with a corresponding reduction in the energy of the system) by the tunnelling of a full one-electron charge, this process can take place. In some circumstances of low constant-current flow the potential oscillates as individual electrons tunnel alternately backwards and forwards across the interface. The oscillation frequency equals the ratio of the current to the charge carried by one electron[24].

Tunnelling is not restricted to electrons. Radioactive decay[25] can also be modelled. The nucleus is in a stable situation by the combination of repulsive and attractive interactions forming an energy well, with particles able to tunnel through the energy barrier and escaping from time to time. We express the escape probability using the wave properties of the escaping particle.

7.6 Delocalisation

In a solid material, the outer or bonding electrons[26] of each atom are 'delocalised', free to move throughout the extent of the material and to occupy specific 'energy bands'. We can model energy bands as quantum states so close together that there is essentially no energy gap between them. (We treat energy bands in much the same way as energies of macroscopic bodies: they have quantum states with negligible separation.)

Metal atoms have only small numbers of outer, bonding electrons and under the influence of an electrical potential these electrons can move readily through relatively unoccupied bands. Consequently, metals are electrical conductors in three dimensions. Graphite, which also conducts electricity, consists of stacked sheets of carbon atoms that provide electron mobility in two dimensions[27].

In contrast, the atoms of electrical insulators have more outer or bonding electrons in closed or filled bands and these are not free to move and carry current. Semiconductors exhibit intermediate behaviour. When provided with additional energy, some bonding electrons are promoted to a higher-level band, the conduction band (figure 7.1). The amount of energy required to trigger conduction (the band-gap energy) varies from one material to another, and to some extent this is under our control by 'doping': selectively introducing into a pure material some foreign atoms with different properties. In some situations it is convenient to model electrons as penetrating the solid in a gas-like manner. In an extension of this idea it is possible to devise a two-dimensional 'electron gas' by trapping electrons in a

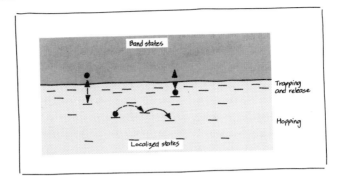

Figure 7.1 Energy levels in an amorphous semiconductor. Electron transport occurs readily in the band states, but there is also transport by hopping transitions between localised energy states. After Scher *et al* (1991).

horizontal plane in a suitably layered semiconductor. In the right circumstances electrons have relatively long collision-free paths, resulting in 'ballistic electron' or 'high-electron-mobility' transistors[28].

'Electrical resistance' usually accompanies the transport of electrons. We model it as a kind of 'friction' causing electrons in a current to collide and dissipate energy as heat. This is similar to the kinetic energy in a flowing fluid being dissipated as heat through viscous effects. Superconductors have no resistance, and so can sustain a direct current without loss or generation of heat. Some materials can be either conductors or superconductors, with particular transition temperatures below which they become superconducting. Electric currents and magnetic fields, as well as pressure, affect the transition temperature value. Particularly important are the 'high-temperature' superconductors, which require cooling only by inexpensive liquid nitrogen (above 77 kelvin or minus 196 degrees Celsius) rather than by liquid helium[29].

The description of electrical conduction in terms familiar to us from our everyday experience requires us to draw on a variety of models. In a classical or macroscopic model, electrical conduction occurs by the acceleration of electrons in an electric field. This provides a slow 'drift velocity' (typically a few **millimetres** per second) superimposed on their much greater random velocities (a few **hundred kilometres** per second). The 'elastic mean free path' is a measure of the distance the electrons travel between collisions and therefore a measure of how good they are as conductors. Now for the wave model. In a good metallic conductor like copper at room temperature this mean free path is a few **tens** of nanometres. The wavelength of the electron (which is comparable to the atomic spacing in the crystal) is a few **tenths** of a nanometre. Because the mean free path of the electron as a particle is much longer than the electron wavelength, the classical model works fairly well.

However, in a semiconductor the electron momentum is smaller and so the wavelength is longer. As a result, physicists can make small quantum devices in which the mean free path is comparable to the electron wavelength. If they reduce the effect of collisions by having high-quality crystals (free of defects) and by using low temperatures, the electrons can move relatively long distances without collisions. They generate 'ballistic electrons' with trajectories that are controlled with electric fields. In those semiconductors in which the dimensions are comparable to the electron wavelength, and without electron collisions, the electron motion is one dimensional. The conductance (the ease with which electrons can flow) depends only on fundamental constants and not on the particular properties of the material or the length of the wire. The relevant fundamental constants are Planck's constant and the electron charge value. The conductance is quantised, increasing stepwise as the width of the conductor increases by an electron half-wavelength, corresponding to each additional electron standing wave that can independently propagate charge. (It is interesting from a modelling point of view that we use a standing wave to model an electron which is also a ballistic particle.)

In such a conductor there is nothing to impede electron flow but, as in all quantum effects, measurement perturbs the system: collisions occur at the contacts with the measuring circuit.

7.7 Quantum Dots

Electrons existing freely within a bulk solid semiconductor have access to effectively continuous energy bands when we apply an electrical potential. It is possible to make very thin, nanometre-scale[30] layers of semiconductors, layers that are thinner than the characteristic wavelength of electrons. This effectively restricts electrons to two dimensions in these 'semiconductor quantum heterostructures'. A narrow strip of such a layer provides one-dimensional containment and a small portion of such a strip would provide a zero-dimensional trap. There are also electrically conducting polymers that provide a one-dimensional path for electrons. We can use a scanning tunnelling microscope to charge a single liquid crystal molecule with a single electron in this one-dimensional situation. In this way we are using macroscopic objects to reveal the behaviour of quantum mechanical objects in a variety of novel situations[31].

Confinement of electrons in a two-dimensional plane causes the energy quantisation to become apparent: a 'quantum well' into which electrons can flow, analogous to water flowing into a well. Rather than confining them by physical barriers, physicists trap electrons in 'low-energy' areas of a semiconducting material. They do this by sandwiching a thin, low-band-gap semiconductor between layers of a higher-band-gap material or on the surface of a metal crystal. The adjacent material acts as two-dimensional quantum barriers or 'hills'. Energy states are even more sharply defined in a 'quantum wire'. The ultimate is the 'quantum dot' where the electron motion is restricted and quantised in all directions. These are 'artificial atoms' or 'model atoms' just as geonium is an artificial atom. They confine the electrons by an applied voltage in a 'box' or 'well' (made up of thousands or up to hundreds of thousands of atoms). The electrons occupy discrete energy levels as they would in a real atom, but scientists can vary these levels by changing the conditions (more easily than is possible in a real atom). Chemists have made clusters of atoms of similar size for other purposes, for example crystallites of cadmium selenide with colours ranging from yellow to red. In another approach, careful etching of silicon with hydrofluoric acid produces pores of nanometre diameter but with lengths of the order of microns (thousands of nanometres). If the pore density is high enough we are left with an array of quantum wires[32].

The properties of the two-dimensionally confined electrons vary depending on their electromagnetic environment. If we wish, we can discuss the properties of these 'electrons in flatland' as new states of matter[33]. These new states differ as fundamentally from each other as steam differs from water or ice, but of course they are not so readily available or accessible.

We can describe the effect of confining an electron within traps of two, one or zero dimensions in terms of the density of electron states. We express the 'occupancy' of specific energy states or quantum levels of an electron as a function of energy (figure 7.2). In an unconfined sample of semiconducting material there is a smooth distribution of density of states with energy, the density increasing with energy, like a **hillside**[34]. When we confine an electron to a two-dimensional layer of the same semiconductor with a thickness comparable to the electron wavelength, it exhibits a distribution graph in the shape of **steps** rather than a hill. The graph for a quantum wire in one dimension has a series of **peaks** and that for a zero-dimensional quantum dot confined in all three dimensions is completely discontinuous with well separated **spikes**.

There is another thing that happens when we closely confine electrons so they demonstrate wave-like behaviour: their tunnelling ability increases. We can visualise this 'resonance' effect on our simple water model as waves being reflected off the walls of a small container and the water splashing out. Therefore current enhancement is possible as a result of increased density of states at a particular energy level and increased tunnelling[35]. 'Josephson junctions' (due to Brian Josephson, then at Cambridge University) arise when tunnel junctions of superconductors are separated by a thin insulating layer. The pairs of electrons[36] that carry the supercurrent can tunnel through the barrier, so a current can flow without an electrical potential up to a certain critical value. Quantum-effect devices in the future may take advantage of the very phenomena that are starting to cause problems as we make transistors smaller and smaller. One of these has the very

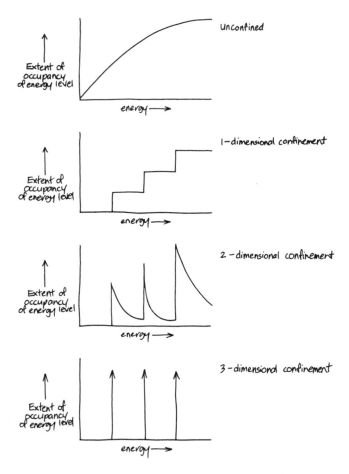

Figure 7.2 The effect of confining an electron within traps of two, one or zero dimensions can be represented in terms of the occupancy of specific energy states of an electron as a function of energy. In an unconfined sample of semiconducting material there is a smooth distribution of density of states with energy, the density increasing with energy, like a hillside. When confined in one dimension to a two-dimensional layer of height comparable to the wavelength of an electron, the same semiconducting material in the form of a quantum well exhibits an electron density of states graph in the shape of steps rather than a hillside. For a quantum wire in one dimension the graph has a series of peaks and for a zero-dimensional quantum dot confined in all three dimensions the graph is completely discontinuous with well separated spikes.

appealing acronym of SQUID: **S**uperconducting **Qu**antum **I**nterference **D**evices, with Josephson junctions in a loop of superconducting wire. These are extremely sensitive to changes in electromagnetic fields[37]. It now appears possible that we can form Josephson junctions within single superconducting crystals[38].

Shining light onto a semiconductor can induce a flow of electrons, a 'photocurrent', and applying an electrical potential to induce a current flow can cause the emission of photons, 'electroluminescence'[39]. Alternate thin layers of a semiconducting material and a more insulating material can provide an 'avalanche photodetector'. This can successively amplify the electrons generated by a single photon so that we can detect it[40]. It is even possible to build around a wide quantum well in a semiconductor a stack of alternating layers that provide narrow quantum wells to serve as mirrors, separated by one-half of the electron wavelength. The resulting resonance creates conditions such that an electron is trapped in a 'bound state' **above** the well at low temperatures. It is as if water can rest not only in the bottom of an energy well but also floating in mid-air above the well. This is a situation that we cannot envisage from everyday experience with macroscopic objects but which is acceptable behaviour for quantum particles. (If we want to stretch our water well model to the limit, we can think of the water vapour in the air above the well.) This possibility was foreshadowed from quantum mechanical mathematical models as early as 1929, and is now being realised[41]. Once again we must revise our opinions of what to expect of matter, and not restrict our models to those based on practical experience with real, macroscopic materials.

7.8 Colourful Quarks

Let us now turn to some of the 'fundamental' particles of matter. Michael Faraday combined or 'unified' the models of electricity and magnetism. Modern physicists are still trying to develop a model that 'unifies' all our models into one 'grand unified theory' or 'GUT'. Although we may eventually do this conceptually with a new paradigm or comprehensive model, the experimental test of

complete unification models by means of high-energy accelerators is beyond our capabilities by many orders of magnitude[42]. Of course, this is no reason to stop high-energy experiments.

The current model assumes four fundamental interactions in nature:

- electromagnetism,

- the nuclear strong interaction,

- the nuclear weak interaction, and

- (by far the weakest) gravitation.

Quarks within neutrons and protons experience a strong nuclear interaction, binding them into nuclei despite the electrostatic repulsion of the like-charged protons. Particle physicists describe this in terms of a 'colour field' quantised into gluons. Electrons do not experience this interaction, and remain free of the nuclei in atoms. Protons and neutrons belong to the large group of experimentally observable subatomic particles that experience the strong nuclear interactions but which are not elementary or fundamental particles. They are 'hadrons', from the Greek word *hadros* meaning strong or robust. Each proton and each neutron is made up of quarks. All other hadrons have greater mass than protons and neutrons and decay rapidly to them. Quarks experience strong, weak and electromagnetic interactions, while the other class of 'fundamental' particle, the 'leptons' (including electrons) experience only weak and electromagnetic interactions.

The 'Standard Model', based on quantum field models (the integration of general relativity with quantum mechanics) includes twelve 'matter particles' with spin one-half. There are six quarks and six 'leptons'[43]. Leptons (a name meaning 'light thing') are those particles that experience the weak interaction but not the strong interaction, and can exist in the free state. They include the electron, 'muon', 'tau' and three varieties of 'neutrino'. There are six varieties or 'flavours' of quarks: 'up' and 'down', 'strange' and 'charm', and 'bottom' (or 'beauty') and 'top' (or 'truth'). Free quarks would have electric charges of plus or minus one-third or two-thirds of an electron charge, but the model requires them to occur only in groups of two or three. The total charge of a group is therefore either zero or an integral number of electron charges. Protons and neutrons are well known groupings, involving only 'up' quarks and 'down' quarks.

- A proton comprises two up quarks and a down quark (charges $+2/3$, $+2/3$, $-1/3$, making a total charge of $+1$).

- A neutron comprises two down quarks and an up quark (charges $-1/3$, $-1/3$, $+2/3$, making zero charge).

(Feynman has noted[44] that if Benjamin Franklin had known about quarks, he would have made the charge of an electron minus three, so that quarks would have unit charges!) There is some experimental evidence for the 'top' quark, which had been the only 'missing' particle from the set required by the Standard Model.

Associated with these matter particles are the field quanta or force-carrying particles. We can classify these into four groups by the strength and nature of the force they carry, and by the particles with which they interact. They have integral spin characteristics (0, 1 or 2). The best known of these is the spin-one particle, the photon, the agent of the electromagnetic interaction or electromagnetic field. Electrically charged particles like electrons repel and attract through an electromagnetic field. In a similar way, quarks interact because of a strong interaction modelled as the colour field and its field quanta, gluons (also spin-one particles). The force-carrying particles have a definite rest mass, which is zero for photons.

Quarks were 'invented' by Murray Gell-Mann and George Zweig of Caltech in 1964. They were part of a comprehensive model proposed to rationalise the growing number of hadrons required to correlate experimental electron scattering results. A convenient model for a hadron is a 'bag' in which quarks move freely but from which they cannot escape. Quarks inside the hadron bag can change their identity in an operation involving the weak interaction, such as beta radioactive decay of a neutron to a proton.

This model, not generally accepted for ten years, did for particle physics what the chemical periodic table[45] had done for chemistry a hundred years earlier. It provided a 'periodic table' of hadrons, and shared with the chemical periodicity model the successful prediction of 'missing' components. Quarks now have amongst scientists a perceived status as 'real' particles almost equivalent to those of protons and neutrons and electrons. Debate has now turned to whether quarks themselves contain even smaller components, such as loops of 'string' in 'string theory'. The name 'quark' apparently originated in the phrase 'three quarks' used by James Joyce in his novel *Finnegans Wake*[46].

Colour (red, green, blue) can be thought of as a kind of charge through which quarks interact. Clearly the names 'colour' and 'flavour' are simply aspects of models that assist us to visualise and accept phenomena totally beyond our experience. We should not expect attributes of 'ordinary' flavour or colour. Quarks are, of course, much smaller than the wavelength of visible electromagnetic radiation so cannot interact with it to provide any impression of physical 'colour'. The idea of 'colour' gave rise to the name 'quantum chromodynamics' or QCD.

Let us put these concepts of colour and flavour into perspective. Up to now I have talked about 'positive' protons and 'negative' electrons without bothering to describe what I mean by 'positive' and 'negative'. From everyday experience we have grown to accept that electrical charges come in two kinds (which we choose to call 'positive' and 'negative') such that a unit positive charge will cancel out a unit negative charge. We take for granted the plus and minus signs on opposite ends of batteries. This idea is now part of our composite model of the way things work, and we don't usually question it. Presumably we use the words 'positive' and 'negative' for the two terminals of a battery and the two kinds of electric charge because in ordinary language they represent pairs of words such as 'favourable' and 'unfavourable' or 'yes' and 'no'. They are terms for quantities that cancel each other out. We could just as easily have named the electric charges 'up' and 'down', and if we had used those words for that purpose they would not have been available to describe the flavours of quarks.

The reason for using 'colour' as a word-model for the properties of quarks is that the strong nuclear interaction only binds particles into combinations that have no colour. By analogy with a combination of coloured lights making white light, such a 'triplet' of quarks is stable, forming a proton or neutron. (A pair of up and down quarks of the same colour can also exist. An up quark cancels out a down quark of the same colour just as a positive charge cancels out a negative charge.)

The weak interaction acts on all matter particles with a spin of one-half, but not on particles of integral spin such as the photon. Neutrinos experience only weak interactions as they have no electric charge or magnetic moment. The weak interaction and the electromagnetic interaction are interdependent (when these interactions are combined or 'unified' they make up the 'electroweak' interaction), and the 'grand unified theory' attempts to combine these two interactions with the strong interaction. We ignore the fourth interaction, the gravitational interaction and its proposed field particle the graviton, in this model. Any 'ultimate' model of matter must involve gravitation also[47]. Gravitation is extremely weak, but it is long range and is always attractive so the effect is cumulative and becomes significant with large masses like planets. This is unlike the other three interactions, which are either short range or have cancelling attractive and repulsive effects. (We should remind ourselves regularly that the current classification of interactions is just another model, and concentrating on it may distract us from seeing a deeper truth.)

In the Standard Model of particle physics, neutrinos are massless, but there is recent evidence that neutrinos are superpositions of states, oscillating among the three varieties. This implies they have some mass, which would require an extension of the model[48]. Neutrinos are probably the key to further advances, and physicists are anticipating an upgraded model of particles and interactions.

In this current model of atomic and subatomic interactions, the force-carrying particles (with integral spins, zero, one or two) carry the interactions between matter particles (which have spin one-half). Calling on our familiarity with classical mechanics, we can visualise a matter particle like an electron or a quark emitting a force-carrying particle that interacts with and is absorbed by another matter particle. Both emitting and absorbing matter particles experience a change in velocity due to transfer of momentum by the force-carrying particle.

Because of their spin properties, the force particles are bosons and the matter particles are fermions. If it were not for the exclusion phenomenon described by the Pauli principle for fermions, the integral-spin force-carrying particles would cause the half-integral-spin matter particles to condense to very high-density states of low energy, rather than forming separate protons and neutrons. Atoms, and the larger objects making up matter as we know it, could not exist.

It is debatable to what extent we should use 'ordinary' classical or everyday phenomena to provide very much simplified models of subatomic interactions[49]. It is worth pointing out the model of two ice-skaters transferring momentum without contact between them by throwing heavy objects to each other. They can also transfer angular momentum if the object thrown is a boomerang!

Finer details of quantum field models provide some further satisfactory correlations, but also show up inadequacies in the current model which are still being studied. These are particularly associated with the gravitational properties of the 'antimatter' particles. The first example was the positively charged positron (antielectron) first identified in 1932[50]. Bubble chambers reveal the process of materialisation of high-energy photons (gamma rays) into electron–positron pairs[51]. Every matter particle is now modelled as having a matching antiparticle with which it can annihilate. The same force-carrying particles are appropriate to both particles and antiparticles[52].

There is also the search for 'strange matter': neutron matter of atomic mass[53] greater than the heaviest atomic nucleus but far less than that of the very dense neutron stars or the unimaginably dense black holes. It has been proposed that the 'strange quark' may be involved[54].

7.9 Virtual Reality

An interesting model introduced by Paul Dirac and useful for discussing antimatter is that of negative energy. This idea arises naturally from mathematical models (which often provide 'negative' solutions as well as the 'positive' solutions expected for 'normal' systems). A negative-energy electron provided with sufficient energy from interaction with a high-energy photon would become an ordinary, positive-energy electron that would appear to us to have materialised out of empty space. (A negative-energy electron would not be observable in 'our' positive-energy space.) However, the **absence** of a negative-energy, negatively charged particle would be a hole in the negative-energy space and so appear to us as a positive-energy, positively charged particle (a positron). This model conforms with the observed materialisation of electron–positron pairs from high-energy photons. Combining this particle-pair creation with the uncertainty model providing energy 'loans' to quantum particles (the larger the amount, the shorter the term) leads to the idea of 'virtual', short-lived particles occupying all regions of space. Virtual particle models can describe the observed nature of electromagnetic interactions just as well as field models[55].

7.10 Isotopes

For the middle third of the twentieth century most scientists used a model which treated protons and neutrons (collectively named 'nucleons') as fundamental particles within the atomic nucleus. The two particles were much the same except that one had a positive charge matching the electronic negative charge. If a neutron is freed from a nucleus it breaks up into a proton, an electron and a neutrino within a few minutes. The inhibition of this decay within a nucleus fits the Pauli exclusion principle. The proton that would result could not occupy the same quantum state as a proton already in the nucleus. Recall that in the current particle physics model each proton or neutron is made up of three quarks. There is one of each colour, with protons having two up quarks and one down with a resulting unit positive charge, and neutrons having two down quarks and one up quark[56]. Attempts have been made to describe the 'structure' of the nucleus with models analogous to gases, liquids and crystalline solids, although these models may have little to add to our understanding of the properties of the fundamental particles other than providing reassurance[57].

Nevertheless, most chemists and biochemists and geochemists, who deal primarily with interactions between atoms and with structures based on molecules, need look no further into the 'structure' of the nucleus than protons and neutrons. They avoid using more complex or detailed models than necessary: they cannot worry about everything at once, and if they are thinking about biochemical metabolic pathways they do not have time to think about quarks.

In any atom the numbers of protons and neutrons are approximately equal, although considerable variation is possible in the number of neutrons:

- The number of protons in a neutral atom is its 'atomic number', different for each chemical element, and controlling its chemical properties. It determines the number of electrons, and therefore how the atom interacts chemically with other atoms.

- The total number of nucleons (protons plus neutrons) in an atom determines its 'atomic mass' (previously called the 'atomic weight').

- For each possible value of the atomic number (determined by the number of protons) there is more than one possible value of the atomic mass (determined by the number of neutrons), or more than one form for each chemical element.

Each combination of atomic number and atomic mass is a unique nuclear species or 'nuclide'. There are just over a hundred characterised chemical elements, but more than fourteen hundred identified nuclides. 'Isotopes' are pairs (or groups) of nuclides that are chemically almost identical but which have different masses[58].

For example, carbon (atomic number six) exists as three isotopes:

- The most common is carbon-12 with 12 nucleons (6 protons, 6 neutrons).

- This is mixed with some of the form having 13 nucleons (6 protons, 7 neutrons).

- There is also a relatively small amount of carbon with 14 nucleons (6 protons, 8 neutrons).

Some nuclides are unstable (or 'radioactive'), transforming spontaneously and randomly ('decaying') into other nuclides and emitting high-energy helium nuclei (alpha particles), electrons (beta particles) and electromagnetic gamma rays[59]. These decay processes for unstable nuclides can occur in fractions of a second or take millions of years. Isotopes, as well as having different masses, therefore sometimes have different radioactive decay processes. In popular use, the word 'isotope' implies 'radioactive', but this is not so in scientific use. There are 'stable isotopes' as well as 'radioactive isotopes'.

If we visualise the collection of nucleons (neutrons and protons) in an atomic nucleus as a liquid droplet[60] we have to extend the model for some nuclei which are **particularly** unstable. Neutrons and protons tend to bind together in pairs (as 'deuterons') that attract each other and other neutrons. Nuclei with unequal numbers of neutrons and protons exist with varying degrees of stability. On the very limits of stability there are unusual nuclei (such as lithium-11 with 3 protons and 8 neutrons) which are larger than ordinary nuclei with about the same numbers of nucleons. If we model ordinary nuclei as liquid droplets, we can think of these very unstable nuclei as being surrounded by a misty cloud or 'halo'. Tunnelling permits two neutrons in the lithium-11 nucleus to roam outside the liquid droplet.

Endnotes

[1]Postle (1976, Chapter 5) expresses this in terms of the need to 'know-it-all-at-once'.

[2]A gas is called a 'vapour' at temperatures below the critical temperature for that material.

[3]Fortman (1993).

[4]See **11.3 Superfluid Helium**.

[5]Further reading Postle (1976, Chapter 5). See also **3.4 Fields**.

[6]See **9.10 Microclusters**.

[7]See **5.8 Phase Changes**.

[8]Further reading Jones (1992), Barlow (1992), Goldman *et al* (1988), Postle (1976, Chapter 8). See also **5.8 Phase Changes** and **13.3 Nucleation**.

[9]See **8.1 Atom Models**.

[10]The electron charge is 1.6×10^{-19} coulomb.

[11]See **7.5 Tunnelling**.

[12]Likharev and Claeson (1992).

[13]However, we may be asking ourselves why we use the names 'positive' and 'negative' for electric charges: see **7.8 Colourful Quarks**.

[14]Further reading Maddox (1992a), Postle (1976, Chapter 9), Kastner (1993, p 25). See also **4.6 Quantum Leap** and **7.8 Colourful Quarks**.

[15]See **4.2 Wave States**.

[16]Further reading Reed (1993). See also **4.5 Playing Cards and Four-Strokes**, **4.2 Wave States** and **4.3 Wave in a Box**.

[17]The electrons used for this purpose typically have a wavelength of 10^{-15} metre. Further reading Ryder (1992).

[18]Further reading Ekstrom and Wineland (1980), Kastner (1993).

[19]Devices for counting and transferring single electrons have also been described (further reading Corcoran (1990, p 82), Nejoh (1991)) and other techniques for confining electrons are discussed in **7.7 Quantum Dots**.

[20]See **4.1 Ultraviolet Catastrophe**.

[21]Further reading Austin and Bertsch (1995).

[22]Further reading Davies (1980, Chapter 4).

[23]Further reading Bate (1988), Likharev and Claeson (1992).

[24]See also **7.3 Charge Integrity**. Further reading Likharev and Claeson (1992).

Similar effects with semiconductors (**7.6 Delocalisation** and Chapter 13) provide the basis for a variety of technological devices (**7.7 Quantum Dots**).

[25]See **7.10 Isotopes**.

[26]See **9.3 Bond Models**.

[27]There are also organic conductors (**13.5 Organic Conductors**) that confine conducting electrons in a similar way.

[28]See below. Further reading Corcoran (1990). There are various other materials not normally considered to be metals that have metallic or semimetallic properties, such as potassium-doped C_{60} (**13.5 Organic Conductors**).

[29]See **13.7 Superconductors**.

[30]Of the order of ten nanometres; atomic diameters are typically tenths of a nanometre, see **13.1 Layer by Layer**.

[31]Scanning microscopy is discussed in **8.10 Surface Images**. Further reading Corcoran (1990), Eaves (1992), Nejoh (1991), Reed (1993), Canham (1993), Kastner (1993), Chang and Esaki (1992). See also **13.5 Organic Conductors**.

[32]Methods of constructing these semiconductor electron containment devices are explained by Corcoran (1990), Main (1993), Canham (1993), Kastner (1993) and Reed (1993) and trapping of electrons on a metal surface is discussed by Zettl (1993), Wilkinson *et al* (1996) and Heller (1996). See also **9.10 Microclusters**.

[33]Further reading Kivelson *et al* (1996).

[34]Bate (1988, p 81), Corcoran (1990, p 46), Chang and Esaki (1992, p 40).

[35]Compare **4.3 Wave in a Box**. Further reading Chang and Esaki (1992).

[36]The pairs of electrons are called Cooper pairs. See **13.7 Superconductors**.

[37]Further reading Wolsky *et al* (1989), Rae (1986, p 89).

[38]Further reading Bate (1988), Pegrum (1992).

[39]Further reading Canham (1993).

[40]Further reading Morgan (1993).

[41]Further reading Washburn (1992), Capasso *et al* (1992).

[42]Further reading Horgan (1995, 1996), Cline (1994). For 'paradigm', see **1.5 Hindsight and Foresight**. See also **1.3 Galloping Horses** and **3.4 Fields**.

[43]See **4.5 Playing Cards and Four-Strokes**. Further reading Ryder (1992), Adams (1992), Hawking (1988, p 67), Riordan (1992), Lambourne (1992), Barlow (1992), Davies (1985), Sutton (1993), Crawford and Greiner (1994), Cline (1994).

[44]Feynman (1985, Chapter 4).

[45]See **8.6 Periodic Elements**.

[46]Further reading Riordan (1992), Farmelo (1992), Hawking (1988, p 69).

[47]Further reading Ryder (1992), Goldman *et al* (1988).

[48]Further reading Flam (1992), Adams (1992).

[49]See also **4.5 Playing Cards and Four-Strokes**. Further reading Farmelo (1992), Postle (1976, Chapter 8).

[50]Further reading Goldman *et al* (1988).

[51]Further reading Jones (1992).

[52]The chirality or 'handedness' of elementary particles is introduced in **9.12 Handedness**.

[53]See **7.10 Isotopes**.

[54]Further reading Crawford and Greiner (1994).

[55]See **3.4 Fields**. Further reading Davies (1980, Chapter 4).

[56]See **7.8 Colourful Quarks**. A still earlier model (**8.1 Atom Models**) considered the nucleus to be made up of electrons and a (different) number of fundamental units that were called 'protons', a word derived from the Greek for 'first'. Further reading Hawking (1988, p 68).

[57]Further reading Cook and Dallacasa (1988).

[58]The words 'nuclide' and 'isotope' have slightly different meanings, the former referring to the description of a particular nuclear species and the latter applying to the nuclides of a particular **chemical element**. A 'chemical element' is often called an 'element' if there is no possible confusion with other kinds of 'element'.

[59]Stewart (1993) and Buck *et al* (1993) deal with the distribution of alpha decay half-lives.

[60]Further reading Austin and Bertsch (1995).

CHAPTER 8

ATOMIC MATTER

The atom is the reference point for the whole of the science of materials, whatever we choose to call the particular discipline: physics, chemistry, geology, biochemistry, and so on. We have had the atom as a model for so long, and we have tested the model in so many ways, that we have accepted it as reality. We have done this despite any reservations we may have about the current models of some of its more fundamental component parts.

8.1 Atom Models

William Higgins, John Dalton and other scientists in the late seventeenth century developed a model that satisfactorily described two basic experimental observations during chemical reactions:

- conservation of material, and

- participation by different materials in definite proportions.

The atomic model as formulated by Dalton proposes that the chemical elements are composed of indivisible atoms, all the atoms of a given element possessing identical properties. (To be precise, atoms can have slightly different masses and therefore not **exactly** the same chemical properties: the idea of isotopes described at the end of the previous chapter did not come until later. However, the **chemical** differences between isotopes are small.)

The atoms of different elements have properties that are significantly different from each other. Chemical change or chemical reaction involves the combination, separation or rearrangement of atoms, without any change in the atoms themselves. Because atoms in this model are indivisible, they combine in fixed ratios when they react to form molecules, as seen in the next chapter. Any particular pure compound contains only one kind of molecule, and that molecule always contains the same whole number of atoms of each element. In the atomic model of matter, mass exists in discrete amounts: it is **quantised**.

While more recent models have refined our perceptions of some of these assumptions, the atomic model has survived unchanged to the present time as the basis of chemistry. The atom is one of our most successful models, and for this reason we no longer regard it as a theory or a model, but as reality[1]†.

J J Thomson in the first years of the twentieth century proposed as a model for the atom a sphere of positive electricity in which were embedded enough electrons to neutralise the positive charge. Because the mass of an electron is so much less than that of an atom, this model required most of the mass to be associated with the positive charge. This 'Thomson model' of an atom with a positively charged and more-or-less uniformly massive nucleus was very short-lived, because of an experiment conducted a few years later. Ernest Rutherford, Hans Geiger and Ernest Marsden observed that most of the positively charged alpha particles striking a thin metal foil were undeflected or only slightly deflected, but that a few bounced back, reversed in direction. The resulting 'Rutherford model' of the atom in line with these observations was an atom that was mainly 'empty space' with the positive charge concentrated in a very small, dense nucleus. The number of positive charges was equal to the atomic number of the element.

Early in the seventeenth century, Johannes Kepler had developed a mathematical model that successfully described the motion of the planets. Astronomers describe the shapes of planetary elliptical orbits by the lengths of major and minor

† Endnotes for this chapter can be found on page 189.

diameters, but the period of motion (the length of the planet's year) depends only on the major diameter. Later, Isaac Newton modelled planetary motions by gravitational attractions that were proportional to the masses and inversely proportional to the square of their distance apart[2].

After Ernest Rutherford had shown by using alpha particles as probes that atoms have an internal structure with a very small dense nucleus, the model for the atom changed. Electrons in planetary orbits around the nucleus in this model electrostatically attracted the positive nucleus. The interaction was proportional to the square of the separation distance, just as in the gravitational interaction between Sun and planets. In this model of the atom due to Niels Bohr, electrons moved in certain specified elliptical orbits around the nucleus. (This overcame

objections to the classical mechanics model that predicted the electrons would spiral inevitably into the nucleus.) The energy of the electron in orbits with different shapes would be the same, and the physical property corresponding to different elliptical shapes is the angular momentum. The additional assumption in the Bohr atom model not present in the planetary model was the limitation to a finite number of orbital shapes (angular momentum values) for each energy value: quantisation of angular momentum.

8.2 Quantum Mechanical Atom

The Bohr atom is important historically, but in the quantum mechanical model of the atom all the details proposed in the Bohr atom, including quantisation, arise without the need for any assumptions. We now have a choice in modelling the behaviour and energy of an electron in the vicinity of an atomic nucleus:

- we can use a modified classical model (which is easy to visualise), or

- we can use a more sophisticated, self-contained quantum mechanical model, or

- we can combine the most useful aspects of both.

At low energies, the classical and quantum mechanical models of hydrogen and hydrogen-like atoms provide charge density values which are very different (although, as explained shortly, for the Rydberg atom they are similar). A significant difference is that quantum mechanics predicts some charge density beyond the limits of the classical model.

In this 'classical' model, with the electrons as orbiting particles, the heavier elements have electron velocities that are a considerable proportion (more than 50%) of the velocity of light. Therefore these must be 'relativistic' rather than conventional 'classical' calculations[3].

8.3 Rydberg Atom

Highly excited or energetic atoms, called 'Rydberg atoms' after a Swedish spectroscopist, provide a link between classical and quantum physics. A Rydberg atom, an atom with its outermost electron of very high energy, can have a diameter as large as ten micrometres and a lifetime of the order of a second.

Rydberg atoms are like hydrogen atoms in some ways. There is one excited electron of charge minus-one. This is attracted by a compact, massive ionic core of charge plus-one made up of the nucleus plus all the inner electrons.

Conventionally, we designate the energy of the electron as negative when it is associated with the nucleus. This indicates that work must be done to overcome the attractive electrostatic or coulombic interactions between the negative electron and the positive nucleus. As we give increasing amounts of energy to the electron, its energy becomes less negative, until it approaches zero when the electron and proton are well separated and independent of each other, fully ionised.

This model describes the allowed, quantised vales of the energy in terms of the principal quantum number (a positive whole number, one, two, three, etc)[4]. A principal quantum number of one describes the ground state or **lowest possible energy** state of the atom, so the electron of a hydrogen atom in its ground state (principal quantum number one) has its lowest possible energy. The **highest possible energy** (zero, with the electron lost by the nucleus) corresponds to a principal quantum number approaching infinity.

Ordinary chemistry and ordinary atoms are associated with small values of principal quantum numbers. A Rydberg atom is a 'pseudo hydrogen atom', with one electron like hydrogen, but the electron is in a high-energy state (a state with a large principal quantum number). Radio astronomy has observed principal quantum number values as high as several hundred for Rydberg atoms in outer space. The size of an atom depends very sensitively on the principal quantum number, so these Rydberg atoms are very large compared with 'normal' atoms.

It is interesting that Niels Bohr used the idea of highly excited atoms in his original proposal. He assumed classical behaviour, with the frequency of electromagnetic radiation emitted by a Rydberg atom approaching the orbital frequency of the electron, a suggestion subsequently called the 'correspondence principle'[5].

It is instructive to compare the properties of a geonium atom (described in **7.4 Electron in a Box**) and a Rydberg atom from the point of view of classical and quantum mechanical properties. Geonium is a very large atom (being the size of the Earth) yet it has an electron with quantised properties. On the other hand, the Rydberg atom which is much smaller than geonium (although larger than an 'ordinary' atom) has electrons with properties approaching classical behaviour. From these rather unusual and 'extreme' models of atoms we obtain fresh insights into the properties of materials.

8.4 Quantum Chaos

We can model atomic structures by combining a standing wave (a model from wave mechanics) with a periodic orbit (a model from classical mechanics). However, when there is more than one 'orbiting' electron, another phenomenon arises: chaos.

We can make a classically **stable** model of a hydrogen atom (one proton in the nucleus, one orbiting electron) with stable orbits. However, when there are two electrons (as in helium) it is a classically **chaotic** system. At first sight, it would appear that the classical model would be of no value for a chaotic system. Yet the periodic orbits characteristic of classical systems still form useful components of quantum chaos models for helium and larger atoms. We can visualise the situation as unstable periodic orbits embedded in a sea of more common nonperiodic orbits. This is a kind of quantum interference effect, resulting in 'scars' in the wave function. In wave function terms, these are concentrations of probability amplitude, reflecting fluctuations associated with unstable periodic orbits of the corresponding classical system.

Wave function scarring has been demonstrated experimentally in quantum systems, with potential applications in commercial devices[6].

Again we see that we must combine the features of several models to describe atomic scale systems in a manner consistent with real, measurable effects. We cannot completely discard the classical orbital model when introducing the quantum wave function model.

8.5 Building an Atom

We can describe the current model for the electronic structure of atoms most useful to chemists very simply:

- When matter waves are confined to a particular region of space, only certain wavelengths are allowed.

- Electrons possess certain definite values of energy but no intermediate values.

- The 'allowed' energies are described by quantum numbers.

- The emission or absorption of a quantum of energy accompanies a transition between two energy levels, as we saw above in the Rydberg atom.

- Electrons can be promoted to higher energy levels by absorption of photons.

- The description of the chemical properties of atoms (the science of 'chemistry') deals mainly with the situation where all the electrons of each atom are in their lowest available energy states (the ground states).

Consider the hypothetical process of 'building' an atom by adding one electron at a time to a nucleus. The electrons occupy successive electron energy levels or 'shells', identified by the integral (whole number) values of the principal quantum number and corresponding broadly to the seven rows or periods in the periodic table. The electron shells contain energy sub-levels or 'sub-shells', their energy values described by the integral values of the angular momentum quantum number and the magnetic quantum number. (These quantum numbers describe the behaviour of the electron in a magnetic field.) Each electron occupies a diffuse region in space called an 'orbital'. This name replaces the term 'orbit' of the solar system model of the atom, but allows us to retain some of the associated ideas from the classical model. The electron density of this orbital at any place within the atom is proportional to the probability of finding the electron there. A separate set of quantum numbers describes each orbital:

- The principal quantum number specifies the energy shell to which the orbital belongs.

- The angular momentum quantum number defines the 'shape' of an orbital.

- The magnetic quantum number specifies the particular orientation or direction in space.

- There is also a spin quantum number identifying each electron[7].

Each orbital accommodates only a limited number of electrons. As I have indicated, there is an exclusion rule: no two electrons in an atom can occupy the same quantum energy state[8].

Chemists successfully use these 'atomic orbitals' in considering bonding, molecular interactions and chemical properties. Sometimes they draw orbitals with blurred or 'fuzzy' outlines to emphasise that they portray electron probabilities. On other occasions they give them solid outlines to indicate that a particular 'proportion' of the electron lies within that line. To some extent, however, atomic orbitals are oversold as models. Sometimes we are given the impression that isolated atoms are not spherically symmetric but have directional lobes of electron density, all ready for forming chemical bonds. But there is no suggestion in this model that without external influences the appropriate model for an isolated atom is anything other than a spherical electron cloud. We should think of orbitals as an indication of the **potential** for strength and direction of chemical bonding. Similar comments apply in the simpler model of 'electron pair domains' advocated by Ronald Gillespie and coauthors for those starting to study chemistry[9].

This atomic model describes the bulk properties of the chemical elements (such as why mercury is a liquid but gold is a solid at room temperature) as long as we take relativity into account[10].

8.6 Periodic Elements

Chemists can break down all the vast number of materials around us by 'chemical' processes, involving moderate amounts of energy leaving all nuclei intact, to about a hundred substances. These substances which they can decompose no further are chemical 'elements'. The fundamental particle of each element is a particular kind of atom with a unique atomic number determined by the numbers of protons and neutrons it contains[11].

In stars, hydrogen undergoes nuclear fusion reactions to form helium, and these two chemical elements are common in the Universe. Other chemical elements on Earth could not have arisen in this way. Making a heavy element like uranium from hydrogen takes an enormous amount of energy. The origin of elements heavier than iron on Earth we attribute to the 'nucleosynthesis' accompanying the gravitational collapse of a star (a supernova). Analogous to the 'stored energy' we use when burning chemical fuels, we can picture the heavy elements like uranium as having a store of energy now available as 'nuclear fuel' for nuclear reactors[12].

So chemists had a model that allowed them to describe all the materials that we know on Earth in terms of one or more of a hundred or so chemical elements. Even with such a model, the problem facing them in sorting them all out was enormous. The major activity of chemists during the past 150 years has been to classify these elements on the basis of their chemical and physical properties.

We often classify chemical elements into 'metals' and 'nonmetals'. Within metals there are coinage metals (copper, silver, gold),

alkali metals (lithium, sodium, potassium) and so on. In the current atomic model the chemical properties of the elements depend in very complex ways on the properties of the protons and electrons. As a result, our 'chemical reactivity' models or 'rules' of chemical behaviour have many 'exceptions'. The result is that the scientific field of chemistry still appears very confusing to 'outsiders' despite the major advances already made.

The most important model in the whole of chemistry is the chemical periodic table. At first the chemical elements were arranged in order of increasing atomic masses, the atomic mass being a concept developed early in the nineteenth century by John Dalton. John Newlands in the mid 1860s used a musical model, the 'law of octaves' reflecting the similar properties that recurred periodically, and at about the same time Julius Lothar Meyer published a table of elements. However, we credit Dimitrii Ivanovich Mendeleev with formulating in 1869 the basis of the current chemical periodic model of the elements[13]. Similarities and trends in physical and chemical properties occur both in the rows across the periodic table ('chemical periods') and down the columns ('chemical groups'). Not only did this model accommodate many of the observed properties of the elements, but it also permitted the prediction of elements that were then unknown: the best test for any model.

Mendeleev was sufficiently flexible in his approach to recognise that atomic mass order did not always provide the best 'fit' to his model. He did not hesitate to sacrifice this order if necessary, or to leave gaps for 'undiscovered' elements. This approach was justified, as chemists later realised that atomic number (the number of protons in the nucleus) rather than atomic mass provides a much more satisfactory model.

Attempts at improving the chemical periodic table are still being made, such as a three-dimensional periodic table including the energies of bonding electrons, or periodic tables of compounds and molecules instead of chemical elements and atoms[14].

One of the most useful aspects of the models of atomic structure developed during the early twentieth century was the 'explanation' for the general structure of the chemical periodic table. Thus models developed for different aspects of materials were providing mutual support and reinforcement. However, the models do not account for the detailed chemical properties, and to a large extent chemistry remains a descriptive science. It is possible that eventually a more 'basic' model will provide a simpler and more direct foundation for the observed properties of the chemical elements and the compounds formed between elements. It might have to use the properties of the more fundamental particles making up protons, neutrons and electrons, or even use an approach still unimagined. This vision seems a long way off, and in the meantime we must make the best of the relatively crude chemical models we have.

8.7 Noble Gases

One of the most useful features of the chemical periodic table is that it assists us to rationalise chemical reactivity. This is the tendency of atoms to exchange or share electrons with atoms of the same chemical element or other elements and so (in the thermodynamic model) achieve greater stability. One of the groups of the chemical periodic table comprises the noble gases (helium, neon, argon, krypton, xenon and radon). Chemists say they are 'noble' because they do not readily react, and 'gases' because without strong interactions between their atoms or with other atoms they remain uncondensed, even at low temperatures. These elements are sometimes known as 'inert gases' or 'rare gases', but the title 'noble gases' is preferred. This is the most accurate description of their chemical reactivity: like the 'noble metals' they resist undergoing chemical reactions, but they can react in the right circumstances. They exist in an elementary, monatomic (single-atom) state. Our energy model attributes the low reactivities of the noble gases to their initial electronic structures already being low energy. The precise numbers of electrons in these atoms are particularly favourable to stability. We model the chemical reactions undergone by other atoms as attempts to achieve the particular electronic configurations of the noble gases, with chemical bonds formed in the process.

Noble gases provide an example where misleading models delayed the progress of chemistry. Because chemists believed that noble gases could not form any compounds, they did not look very hard for them. Nevertheless, noble gases provided an important clue to the understanding of chemical bonding, as demonstrated by Walther Kossel, Gilbert N Lewis and Irving Langmuir between 1916 and 1919.

8.8 Cooling Atoms

Because their electrons absorb and emit quanta of energy, atoms act as finely tuned transmitters and receivers of electromagnetic radiation of particular frequencies. Each combination of such frequencies creates a characteristic spectrum for the particular type of atom which is valuable in obtaining information about the structure of matter.

Thermal motion of atoms reduces the precision of the information obtained. Each atom in a gas of randomly moving identical atoms provides a slightly modified or 'shifted' spectrum, with the result that the spectral lines broaden. The Doppler shift, similar to the change in pitch heard by a stationary observer listening to the horn of a passing train, occurs when an atom is moving relative to a light source[15].

Cooling is an obvious way to slow down an atom and so reduce these effects. Although an atom slows in collision with cold surfaces or particles, this procedure destroys the isolation of the atom that was the very reason for studying it. So, more conveniently, we slow atoms by interaction with laser light.

Laser cooling is possible if we direct a laser beam, tuned to a frequency lower than a characteristic frequency of the particular kind of atom, into the atomic gas or beam of atoms. Some of the atoms moving **towards** the laser beam slow by momentum transfer as they absorb and re-radiate the photons because of the Doppler shift. Atoms moving **with** the laser beam are less likely to absorb photons (because their Doppler shift is in the opposite direction) so this acceleration effect is relatively small. It takes large numbers of photons to 'cool' each atom, but the net effect is that the average atomic velocity reduces. The light field exerts a 'viscous' restraining interaction sometimes called 'optical molasses'[16].

8.9 Trapping an Atom

Eventually, when laser beams lower the temperature sufficiently, they can trap atoms in a 'web of light'. The wave model sees this as trapping an atom in the trough of each wave, with insufficient energy to escape. Using another analogy, the atoms are in a landscape made up of ridges and valleys. We now have a lattice of atoms separated not just by fractions of a nanometre as they would be in a solid array, but by hundreds of nanometres, a distance equal to the wavelength of light[17]. Alternatively, once neutral atoms cool to effectively zero velocities, they are held within magnetic field atom traps. This is similar to electrically charged particles being held in electromagnetic field traps. The same technique works for neutrons, which have lifetimes of about fifteen minutes[18]. Single atoms are held and detected also in electromagnetic cavities, the 'one-atom maser' using microwaves, and an optical cavity using lasers[19].

One of the most interesting things about fullerenes (next chapter) is that they provide us with the opportunity of fully isolating or 'shrink-wrapping' an atom in a molecular container. The first achievement of this kind, made by the laser vaporisation of a graphite disc impregnated with lanthanum chloride, was the trapping of a lanthanum atom within the C_{60} and C_{70} fullerenes[20].

8.10 Surface Images

Physicists and chemists have invented a great variety of techniques for studying surfaces of solids, and recent developments show that it is possible to obtain images of the individual atoms making up surfaces. They can also obtain detailed information about molecules adsorbed on surfaces.

Electron microscopy can provide some information on surfaces. High-energy electrons tend to penetrate deeply without interacting with the surface, and low-energy electrons are easily deflected by charges and fields. Nevertheless, high-resolution transmission electron microscopy is able to identify arrays of close-packed **molecules**, each containing sixty carbon atoms[21]. For the next level of detail, imaging individual **atoms**, we need another technique.

Scanning probe microscopes operate by mechanically scanning a sharp tip over the sample surface, sensing and monitoring its deflection and converting this information into a model. Scanning tunnelling microscopes apply an electrical potential between the tip and the sample causing electrons to flow (the tunnelling current) even when they are not in contact, as long as their electron clouds overlap. They have the advantage that we need not maintain the sample in a vacuum. We can observe the charging of a single molecule by one electron[22].

One of the most interesting of the scanning probe microscopes is atomic force microscopy because it provides another insight into the interactions between atoms (figure 8.1). If we model the

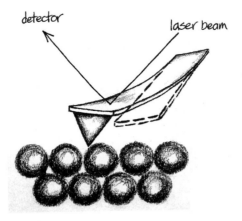

Figure 8.1 Representation of an atomic force microscope. The deflection of the probe (both bending and twisting) is measured by a reflected laser beam.

interaction between atoms as though the bonds were springs, we can evaluate the interatomic spring constant just as if it were an ordinary spring[23]. Although it was not obvious until it was first done, it is surprisingly easy to make a device with a spring constant weaker than a chemical bond 'spring'. The spring constant of a four-millimetre by one-millimetre strip of kitchen aluminium foil is only one-tenth that of an atomic 'spring' constant. On this basis, we can sense atomic-sized displacements without damaging the sample by using a simple cantilever spring bearing a fine enough tip. Optical sensors are very sensitive, with a tip resolution of a hundredth of a nanometre, so the spring constant of the cantilever can be many times greater than that of the interatomic bond. In principle, a sensitive force microscope could even image single electrons[24].

Movement of the tip is sensed by

- electron tunnelling (which requires a conducting surface), or

- an optical detection system such as interferometry or laser beam deflection (so we can make the study under water or in biological environments), or

- magnetic fields associated with magnetic samples.

We can resolve the surface on an atomic scale when the tip is in contact with the surface, providing the loading force on the tip is very small[25]. In these circumstances short-range, repulsive interatomic interactions dominate the interaction, and we can study atomic-scale frictional processes. If, on the other hand, the tip is ten to a hundred nanometres from the sample, longer-range interactions (attractive 'van der Waals interactions', magnetic interactions and electrostatic interactions) are accessible[26].

Scanning tunnelling microscopes (STMs) not only provide images of atoms on surfaces, but are also able to move atoms across surfaces. The repeated transfer of an atom between the microscope tip and the surface is an example of a 'bistable' switch at an atomic level which is significant for advances in microcircuitry[27]. Nanometre-scale recording and erasing for data storage has been demonstrated[28].

8.11 Plasmas

In an extension of the procedure of trapping single atoms by means of electrical and magnetic fields, it is possible to study small groups of ions ('microplasmas') to aid in the investigation of atomic structures. These microplasmas serve as models not only for more complex familiar states of matter and plasmas but also for the far denser one-component plasma (assemblies of ions and electrons) in stellar objects.

One of the features of microplasmas is their low density, with inter-ion distances typically one hundred thousand times those in liquids and solids. Having like charges, the ions mutually repel, maintaining a dispersed structure held together only by the **external** electrical and magnetic fields. This is unlike ordinary condensed matter (liquid or solid) which possesses **internal** attractive interactions[29].

At high temperatures (several thousand degrees Celsius) nuclei liberate their electrons, chemical models cease to be of value, and the material is an array of nuclei and electrons. We can visualise a plasma as a gas-like state of matter, bright and opaque because the particles are freely interacting with photons. (Eventually, at even higher temperatures the particles comprising the nuclei themselves are liberated, but these 'materials' are of little practical importance in human activities.)

The properties of plasmas depend markedly on their densities. In 'tenuous' (low-density) plasmas, like those in neon display tubes, the freed electrons do not exchange energy efficiently with the positive ions and the plasma remains thermally 'cool'. On the other hand, in dense plasmas such as those used in plasma-sprayed coatings of ceramics, the frequent collisions distribute energy more uniformly, so the temperature rises. Also, the greater the opportunity in the plasma at a particular temperature for both

- dissociation of molecules into atoms, and

- ionisation of atoms

the higher the enthalpy or heat content available for applications like spray coating[30].

We believe neutron stars derived from supernovae contain a plasma of iron nuclei and free electrons at extremely high pressure and density. In this situation the iron nuclei behave classically, with like positive charges repelling. However, the electrons are quantum particles, each occupying a different energy state, and because they are forced into very high-energy states they do not interact with the much lower-energy iron nuclei. The result is a one-component plasma of iron nuclei with a uniform background of negative charge. The parameter used to characterise such plasmas is the 'coupling', the ratio of the coulombic potential energy (a function of the density) to the kinetic energy (a function of the temperature). At low couplings a one-component plasma is like a gas, without structure. At high couplings there are liquid-like and eventually crystal-like structures. Our model for the strongly coupled, hot, dense, one-component plasma characteristic of a neutron star is a laboratory system that although diffuse is cold (below one kelvin) and therefore also strongly coupled. The microplasmas described above can serve this purpose[31].

Endnotes

[1]Further reading Kolb (1977).

[2]See **6.1 Interactions**.

[3]See also **4.1 Ultraviolet Catastrophe** and **4.6 Quantum Leap**. Further reading Norrby (1991), Uzer *et al* (1991), Kleppner *et al* (1981).

[4]The allowed energy values of the electron in a hydrogen atom are given by the expression $-E_0/n^2$, where E_0 is a constant with a value of about 13.6 electron volts and n is a positive integer (whole number) called the principal quantum number.

[5]Further reading Kleppner *et al* (1981), Guttzwiller (1992), Haroche and Raimond (1993).

[6]See **3.7 Chaos**. Further reading Heller (1996), Wilkinson *et al* (1996).

[7]For electron spin see **4.4 Heisenberg May Have Been Here!**, **4.5 Playing Cards and Four-Strokes** and **7.8 Colourful Quarks**.

[8]See **4.5 Playing Cards and Four-Strokes**. Further reading Heilbronner and Dunitz (1993, Chapter 10).

[9]See Gillespie *et al* (1996).

[10]See **4.6 Quantum Leap**. Further reading Norrby (1991).

[11]See **7.10 Isotopes**.

[12]Further reading Lovelock (1988), Atkins (1984, Chapter 7).

[13]Further reading Gorin (1996).

[14]Further reading Rouvray (1994).

[15]This concept was introduced in **4.2 Wave States**. There is also a 'relativistic shift' and a 'transit-time broadening' due to the limited observation time for each fast-moving atom.

[16]Further reading Phillips and Metcalf (1987), Chu (1992), Stein (1994).

[17]Further reading Stein (1994).

[18]Further reading Phillips and Metcalf (1987, p 40).

[19]See **4.2 Wave States**. Further reading Baggott (1992, p 29), Haroche and Raimond (1993), Chu (1992).

[20]See **9.14 Fullerenes**. Further reading Curl and Smalley (1991, p 38), Smalley (1991).

[21]See **9.14 Fullerenes**. Further reading Buseck *et al* (1992). As another example, the small 'splats' in plasma coatings (**8.11 Plasmas**) were investigated by scanning electron microscopy for their general shapes and by transmission electron spectroscopy for examination of their internal structures. Further reading Herman (1988).

[22]See **7.5 Tunnelling**. Further reading Binnig and Rohrer (1985), Wickramsinghe (1989), Nejoh (1991), Braun (1992), Welland (1994), Jones and McConville (1995).

[23]The interatomic spring constant is of the order of 10 N m^{-1} and the order of magnitude force required to break a chemical bond (to separate bonded atoms by 0.1 nanometre) is a nanonewton, 10^{-9} N.

[24]Further reading Rugar and Hansma (1990), Salmeron (1993), Overney and Meyer (1993), Welland (1994).

[25]Typically 10^{-7} to 10^{-11} N.

[26]See also **10.1 Molecular Interactions** and **14.14 Nanotribology**. Further reading Rugar and Hansma (1990).

[27]Further reading Eigler *et al* (1991), Quate (1991).

[28]Further reading Sato and Tsukamoto (1993). STMs can also be used to study electronic behaviour on surfaces, including the quantum-mechanical standing waves (with wavelengths of the order of a hundred nanometres) revealing the presence of point defects (further reading Bollinger and Wineland (1990), Curl and Smalley (1991, p 37)), and to image vortices (**13.7 Superconductors**).

[29]Systems such as these, which are too small to behave like macroscopic systems but which are more complex than single atoms or ions, are described as 'mesoscopic', from the Greek meaning 'intermediate'. Further reading Bollinger and Wineland (1990), Curl and Smalley (1991, p 37). See also **11.1 Condensed States** and **8.11 Plasmas**.

[30]Further reading Herman (1988). See also **13.6 Ceramics**.

[31]Further reading Bollinger and Wineland (1990).

CHAPTER 9

CHEMICAL MATTER

The description of the ways in which atoms combine to make molecules and of the properties of these molecules has become known as 'chemistry'. Chemistry has a specialised language and symbolism that tends to discourage those outside (even other scientists) from exploring it too deeply. Nevertheless, it plays such a central role in materials that it is important for everyone to be 'chemically literate'.

9.1 Amount

'Amount' is one of the words that has a more specialised meaning in chemistry than in general usage. When we think about a molecule of water (H_2O) we realise there are two atoms of hydrogen for every atom of oxygen. In other words, there is twice the amount of hydrogen. (This does not mean twice the mass: hydrogen atoms are much lighter than oxygen, so the mass of oxygen is still greater than the mass of hydrogen.) The unit of amount is the 'mole'[1]†, a word derived from the Latin for 'massive heap'. The mole **is** a 'massive heap' of particles, but not a large amount in human terms: there are only about ten moles of water molecules in a drinking glass. 'Amount' as a chemical term is officially recognised, but is still not widely used. Most chemists still prefer to say 'number of moles' rather than 'amount'[2].

† Endnotes for this chapter can be found on page 216.

Associated with the concept of chemical amount is that of 'stoichiometry'. When we form water from its chemical elements, we require twice the amount of hydrogen as of oxygen. A complete description of this reaction is that two molecules of hydrogen (H_2) react with one molecule of oxygen (O_2) to give two molecules of water (H_2O). Thus in water the 'stoichiometric ratio' of hydrogen to oxygen is 2:1[3].

9.2 Molecules

Amedeo Avogadro early in the nineteenth century introduced the term 'molecule', meaning a particle composed of atoms, either similar or dissimilar. Because atoms combine in definite numbers in molecules, it follows that chemical elements combine in definite proportions to make compounds, a compound being a pure substance made up of more than one element.

Some chemical elements when not combined with other elements exist as free atoms rather than as molecules. There is no generally accepted collective term to describe both molecules (like O_2, oxygen gas, or H_2O, water) and free atoms (like O, oxygen atoms), but we are using 'molecular particle' for this purpose. If a molecular particle carries a net electrical charge it is an 'ion' rather than a molecule or atom. 'Cations' carry positive charges and 'anions' carry negative charges.

We consider two or more atoms associated for times which are significant on a human scale to have independent existence as a molecule. Now we have our models of atoms, the next step is to visualise the process of combining them into molecules.

The matter-wave model which has been so successful in 'explaining' the properties of atoms also accounts for chemical bonding and the formation of molecules. Electron waves encompass all nuclei in the molecule, counteracting and balancing the repulsion expected between positive nuclei.

9.3 Bond Models

In the particular conditions existing on the Earth the atom is the most convenient building block for our models of matter, but to the early scientists this was not immediately obvious. To start with, pure materials are rare in nature, most occurring as mixtures with other compounds. It is also rare for chemical elements to exist as isolated atoms. Except for the noble gases (such as helium, neon and argon) they usually combine with other atoms at ordinary temperatures, and to liberate free atoms requires considerable expenditure of energy. Even those chemical elements that occur uncombined with other elements tend to have their atoms chemically bonded to atoms of the same kind. The elements oxygen, hydrogen and nitrogen and the halogens (fluorine, chlorine, bromine) often occur as the 'diatomic' (two-atom) molecules O_2, H_2, F_2, Cl_2 and Br_2. Carbon exists as graphite, diamonds or fullerenes, and gold occurs as a solid metallic crystal. (Mercury is unique as a metal, being almost entirely monatomic in the gas phase[4], although at higher temperatures, above about five thousand degrees Celsius, most elements are in the gaseous, monatomic state.)

It was therefore necessary for early chemists to

- identify pure compounds from amongst natural mixtures, then

- look beyond the pure materials to identify the molecules,

- determine the atomic arrangements within each molecule, and finally

- provide a model of how the atoms combine.

Consideration of the atomic models for the chemically unreactive noble gases eventually provided the electronic configuration basis for an understanding of how atoms interact to form molecules by means of chemical bonds.

When Edward Frankland introduced the term 'bond' in 1866 he stated that he did not intend to convey the idea of any material connection between the component atoms of a compound. He saw the bond more like the gravitational 'bonds' connecting the members of the Solar system. After some initial hesitation, chemists adopted the word-model 'bond', and have found it very convenient.

It would be possible to treat the chemical interactions between atoms in terms of 'interference' effects of overlapping matter waves. However, chemists prefer to treat electrons as probability clouds, with bond formation resulting from the exchange or sharing of electrons by atoms.

The atoms of some chemical elements can acquire a noble gas electronic structure by completely releasing one or more of the outer or bonding electrons. As a result they become positively charged and are positive ions or 'cations'. Other atoms can achieve noble gas structure by gaining electrons and becoming negative ions or 'anions'. The process is 'electron transfer', a transaction in which one kind of atom gains an electron, the other kind loses an electron, and both benefit in terms of stability or energy reduction. The resulting arrays of positive and negative ions form crystalline solid materials described in terms of 'ionic bonds'. This is an extended array of 'bonds' rather than specific interactions between any two atoms. In solutions of these ionic materials, however, individual short-lived 'ion pairs' of oppositely charged ions may occur, temporarily stabilised by the environment provided by the solvent. Other types of atoms 'cooperate' with each other by **sharing** electrons to form molecules by means of 'covalent' bonds, or to form close-packed arrays of atoms with delocalised electrons in metallic bonding. These examples are the extreme or 'ideal' forms of behaviour, and most chemical compounds have bonds with intermediate properties. These bonds are not fully ionic, nor fully covalent, nor fully metallic.

Achievement of noble gas electronic structure for each atom is one aspect of chemical stability. Another is the particular arrangement of atoms within a molecule or as part of an extended network being viewed as 'nature being satisfied' with the structure, or the compounds 'not wanting' to react. The chemical environment and temperature also determine whether or not a structure is stable.

We may analyse types of chemical bonding with the help of the chemical periodic table. For example, the separation into metallic and nonmetallic elements does not appear as part of the basic vertical or horizontal groupings of chemical periodicity. Rather, we see a broad diagonal division running from the top left to the bottom right of the periodic table. The 'most metallic' elements tend to be at the bottom left and the 'least metallic' at the upper right of the periodic table.

While many physical and chemical properties of elements and their compounds correlate well with the chemical periodic table position, the most important are the 'ionisation energy' and the 'electron affinity'. The ionisation energy is the minimum energy required to remove an electron from the ground state of an isolated atom. The smaller the ionisation energy, the easier it is to ionise the atom by removing an electron, and the greater the metallic character of the element[5]. The electron affinity is a measure of the opposite sort of process: the tendency for an electron to be acquired by an isolated atom in its ground state. In general, electron affinity is greatest in the nonmetallic, top-right-hand corner of the periodic table. Combining these two properties of ionisation energies and electron affinities, there is a strong tendency for an atom of a metal to lose an electron to form a positive ion. This electron is transferred to a nonmetallic atom, converting it into a negative ion. As a result an ionic compound is formed. Atoms of elements with intermediate values for the ionisation energy and electron affinity are more likely to **share** electrons in covalent bonds, leading to covalent compounds. All these processes demonstrate the tendency for each atom to acquire noble gas electronic structure, with the thermodynamic model and the energy well model being used to understand the driving force.

Some atoms are able to form chemical bonds with more than one other atom, and the combining capacity (a small whole number) is its 'valence'. ('Valency' is an alternative spelling.) In a molecule of water, H_2O, the valence of the hydrogen atom is one, while that of the oxygen atom is two. As a noun, 'valence' has this quantitative meaning. We also use the word (derived from the Latin for 'capacity') as an adjective meaning 'associated with bonding' in terms such as 'valence electrons' without any quantitative significance. Word-models such as this may arise almost accidentally and pass through periods of rapid change of meaning before 'stabilising' in their finally accepted forms.

9.4 Extended Bonding

When chemical bonds form, they do not always result in separate, discrete molecules (such as the water molecule that contains one oxygen atom and two hydrogen atoms in a stable structure). At the other extreme from this molecular type of structure, there are the (effectively) infinite arrays of covalently bound atoms in covalent network species. These include the three-dimensional structure of carbon atoms in diamond, which is so stable that it breaks down only on melting at 3500 degrees Celsius. (Of course, if oxygen is present, diamond will burn before it melts.) Another example is the alternating array of oppositely charged ions in an ionic solid such as sodium chloride which becomes an ionic liquid only at 800 degrees Celsius. A third type of extended array comprises metal ions and their associated electrons in metals, which also have high melting points. We can model the properties (particularly thermal and electrical conductivities) of these materials as coherence effects of electronic waves over long distances[6].

Intermediate in size between the extremes of the small molecule and the infinite bonded array are various other structures discussed below such as microclusters and macromolecules.

9.5 Classifying Matter

The discussion so far in this chapter has presented the current conventional bonding classification model for substances, with three types of bonding (ionic, covalent and metallic)[7]. The ionic, covalent and metallic models of figure 9.1 (top) represent limiting behaviour.

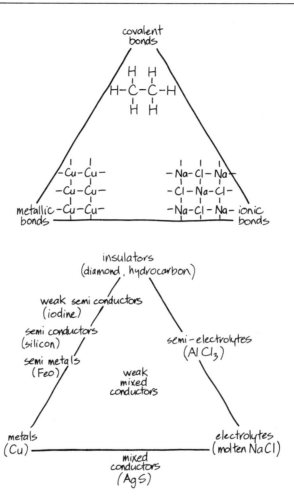

Figure 9.1 Classification of materials. Top: bond model; bottom: electrical conductivity model. (After Nelson (1994).)

Because we cannot directly measure the nature of a chemical bond, it may be preferable to use a classification model based on something we can measure: the value of electrical conductivity[8]. Here the limiting types are 'insulator', 'metal' and 'electrolyte' (figure 9.1 (bottom)).

A metallic conductor (like copper, Cu) has an electrical conductivity that decreases as the temperature increases. This is in contrast to a semiconductor, in which the conductivity increases with increasing temperature. Less familiar are the conducting properties of molten salts (ionic liquids like sodium chloride, NaCl) and of electrolyte solutions (salts dissolved in a liquid such as water) in which the positively and negatively charged ions carry the current. (An interesting model of ionic systems and the electrolyte solutions resulting from their dissolution in water or other similar solvent uses nuts and bolts. We can consider a system with equal numbers of nuts and bolts, nuts representing negative ions and bolts positive ions, for example. In a 'strong' or fully dissociated solution, all the nuts and bolts are disconnected. 'Weak' or partly dissociated solutions can be represented by a system with some of the nuts and bolts screwed together into pairs[9].)

First look at the three corners of the triangle in figure 9.1 (bottom). **Metals** conduct electricity readily in both solid and liquid states by electron flow. **Electrolytes** conduct electricity moderately well in the liquid state by the movement of ions, and sometimes a similar mechanism exists in solids. **Insulators** are poor electrical conductors. It is also convenient to define terms for materials of intermediate character, on the sides of the triangle. **Semiconductors** conduct electricity, but not as well as metals, and the effect of temperature change on electrical conductivity is in the opposite direction. The better conducting semiconductors may be called 'semimetals' or sometimes 'metalloids'. 'Semielectrolytes' conduct electricity in the same way as electrolytes, but not as well. 'Mixed conductors' conduct electricity partly by electron flow like metals and partly by movement of ions. (This classification is independent of the gas–liquid–solid classification: insulators may be solid, liquid or gas, for example.)

Both models in figure 9.1 use terms that describe limiting behaviour. Real materials have bond properties and electrical conductivities on continuously variable scales between these limits. The limiting classifications of both models can be used in similar ways, but they are conceptually different:

ELECTRICAL CONDUCTIVITY	BOND PROPERTY
● electrolyte	● ionic bond
● insulator	● covalent bond
● metal	● metallic bond

Electrical conductivities are physical properties, directly measurable by experiment, while bonds are part of an atomic-scale conceptual model. Experienced chemists use these models (and the associated terminology) interchangeably. Others may prefer to think of an 'insulator' or an 'electrolyte' rather than of a 'covalent bond' or an 'ionic bond'. However, if we want to be able to use all the scientific jargon, we must be familiar with both models.

9.6 Allotropes

Some pure chemical elements exist in more than one form under ordinary temperature or 'ambient' conditions. These are known as 'allotropic' forms. Carbon is one of the best known examples, with the allotropes diamond and graphite having dramatically different properties. Diamond, which is extremely hard, is an electrical insulator and an excellent thermal conductor. It has 'infinite' three-dimensional tetrahedral arrays of equivalent, highly directed carbon–carbon bonds forming a 'cubic' crystal[10]. Graphite, which is relatively soft and is used as a lubricant, is electrically conducting and highly 'anisotropic' (having different atomic arrangements in various directions). It has infinite two-dimensional layers (based on a hexagonal arrangement of carbon–carbon bonds) that interact by weaker forces and slip easily past each other[11].

Until recently chemists had not even contemplated the possible existence of discrete carbon molecules with finite bonding formed by arrays of atoms in hollow, spherical 'soccer balls' or 'capsule' shapes. The molecules were named fullerenes after the structurally strong domes invented by Buckminster Fuller.

Chemists did not really think of fullerenes as allotropes of carbon until they subsequently found that they could produce them relatively easily and in large quantities from graphite and they were abundant in sooty flames[12]. In the solid state, fullerene molecules occur in a crystal lattice of close-packed rotating spheres, totally unlike either diamond or graphite. The transient smaller molecules such as C_2 and C_3 previously identified in carbon vapour, were certainly not treated as allotropes. Chemists had locked themselves into a state of mind requiring allotropic materials to be available in quantities and for times that were significant on a human scale. We should now be more alert to the possible existence of further interesting and potentially useful forms of other elements[13].

Allotropic forms may differ in chemical bonding and molecular composition ('primary allotropy') or less fundamentally differ only in crystal structure. Primary allotropy occurs in the group of elements around the diagonal line in the chemical periodic table that separates metallic and nonmetallic elements. Allotropes of an element tend to differ in the extent of their metallic nature. Diamond is a very good electrical insulator while graphite is a reasonable conductor and has a metallic lustre. Phosphorus occurs as tetrahedral P_4 molecules in the white form and as extensive molecular networks in the red and black forms, the latter with some metallic properties. Sulphur exists in a range of allotropic forms:

- short chains of two, four or six atoms in the vapour,

- cyclic molecules of eight atoms in liquid and crystalline forms, and

- long chains in 'plastic sulphur'.

The allotropic forms of the elements differ in the nature of the bonding between the atoms, and therefore differ in stability: some allotropic forms exist indefinitely in a metastable state. For example, we can grow diamond film by deposition of carbon from a low-pressure hydrocarbon gas. It has no tendency to transform into the thermodynamically more stable graphite[14]. In the case of ozone (O_3) and oxygen (O_2) the stability difference is considerable, but graphite is only slightly more thermo-dynamically stable than diamond at ambient temperature and pressure. We think of tin as a metal, which it is in its 'white' form. However, at low temperatures the nonmetallic 'grey' tin is more stable, and the spontaneous conversion from the metallic white form can cause the collapse of tin objects[15].

9.7 Electronegativity

A covalent bond formed between atoms of the same element (such as the chlorine molecule Cl_2) is nonpolar. The shared electrons are distributed equally between the two nuclei. On the other hand, in an ionic bond the nucleus at one end of the bond has a negative charge, while the other bears a positive charge: an electric dipole. These two models provide the extremes in possible 'bond polarity' between the complete absence of polarity in the pure covalent bond and 'complete' polarity in the pure ionic bond. Between these extremes are 'polar covalent' bonds with all possible degrees of charge separation, the extent of polarity depending on the properties of the particular atoms involved.

The relative tendency of a bonded atom to attract electrons within a bond towards its nucleus is its 'electronegativity'. This name may be misleading at first sight. It does not mean there is necessarily a negative charge on the atom. What is important is the tendency to acquire it. So although the fluorine atom is highly electronegative, once it has acquired an electron and become the fluoride ion, F^-, it no longer attracts electrons. In the chemical periodic table the electronegativity tends to be highest in the top right and lowest in the bottom left, but there is considerable variation in values.

9.8 Giving, Taking and Sharing

The variation in electronegativity values of atoms leads directly to the model that describes those readily giving up electrons as 'electron donors' and others as 'electron acceptors'. While 'giving' and 'taking' electrons they are also 'sharing' them, each atom becoming more like a noble gas atom. This model has proved very useful to chemists.

Chemists have also extended the older idea of oxidation as the reaction of a material with oxygen. We now say that an electron donor species that loses electrons is oxidised. (The even older eighteenth-century model[16] of loss of 'phlogiston' by combustible substances when they burn therefore has some validity if we identify 'phlogiston' with 'electrons'.) Conversely, the species that does the oxidising by gaining electrons and being 'reduced' is the electron acceptor. Magnesium burns vigorously in air, the electron-donating magnesium being oxidised and the electron-accepting oxygen being reduced to an oxide. Similarly, when sodium metal reacts with chlorine to form the salt sodium chloride the electron-donating sodium is oxidised and the electron-accepting chlorine is reduced. Just as the giving and receiving of electrons proceed by interdependent and cooperative mechanisms, so too do oxidation and reduction processes.

9.9 Carbon Compounds

Carbon is one of the elements in the central groups of the chemical periodic table that 'prefers' to share rather than transfer electrons in order to achieve a noble gas electronic structure. Because it has four valence electrons, carbon can form four bonds (called 'sigma bonds'), and in a symmetrical molecule like methane (CH_4) these bonds are symmetrically arranged. If we visualise the carbon atom at the centre of a 'tetrahedron' (a pyramid on a triangular base), then the chemical bonds would point to the corners of the tetrahedron. This tetrahedral model dominates carbon chemistry, with these sigma bonds having electrons localised in the regions between the nuclei. However, in some compounds carbon has fewer than four substituents, with the additional electrons belonging to 'double bonds' (as in 'alkenes' like ethylene) and even 'triple bonds' (as in 'alkynes' like acetylene). For convenience, chemists model each multiple

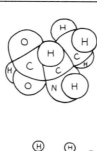

"space-filling"- representation of electron clouds

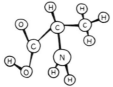

"ball-and-stick" – with location of atom centres and bond orientations

```
 ..         H   H
:O.         ..  ..
 :: C : C : C : H
:O.  ..  ..
 ..  :N:H  H
  H   ..
       H
```

"electron dot" – demonstrating "octet rule", 8 electrons around each carbon, oxygen and nitrogen atom

```
O     H   H
‖     |   |
C  —  C — C — H
 \    |   |
  O   N   H
  |   / \
  H  H   H
```

"two-dimensional" – shows chemical bonds, but not their orientations

$$HOOCCH(NH_2)CH_3$$

"molecular" – identifies the functional groups present

$$C_3H_7NO_2$$

"empirical" – indicates the relative amounts of carbon, hydrogen, oxygen and nitrogen

Figure 9.2 Various ways of representing alanine.

bond as an 'ordinary' single or sigma bond combined with an extra 'pi bond'.

The pi bonds have electrons on either side of the sigma bond separated by a nodal plane (where there is zero electron probability) and not concentrated between the nuclei. A pi bond has been likened to a hot-dog bun, with two halves on either side of the sausage that models the sigma bond and the significance of the nodal plane in the pi bond is often debated[17].

We call compounds of carbon with hydrogen and other chemical elements such as oxygen, sulphur and the halogens 'organic compounds', because they were once thought to be derived from living organisms only. Figure 9.2 presents a variety of methods of representing a molecule of alanine:

● three-dimensional space-filling and ball-and-stick models

● two-dimensional electron-dot and bond representations

● molecular and empirical formulae.

Chemists have built up over the last hundred years an extensive model they call 'organic chemistry', which depicts and facilitates the interconversion of organic compounds. This models electron densities around the carbon skeleton that vary according to the electronegativities of substituents. There is an array of semi-empirical rules and exceptions, procedures and guidelines, which enable synthetic organic chemists to put together molecules of great complexity. This model has sufficient detail to work satisfactorily for most of the time, but is sufficiently simple to master to an adequate level in a reasonable time. The best organic chemists have completely assimilated this model. Almost without conscious thought they can intuitively choose the right reagent and best conditions for obtaining a particular result in one portion of a molecule with minimum disturbance to the rest of the molecule[18]. A simplified model recently introduced[19] uses three types of electronic structure: nonbonding electron pair, shared electron pair (covalent bond) and vacant atomic orbital or electron pair domain. There are five operations that we can perform on these structures, each operation showing (figure 9.3)

- the bond broken
- the bond formed
- arrows representing electron flows
- charge alterations.

This provides a simple example of the application of the organic chemistry model.

9.10 Microclusters

Clusters of tens or hundreds of atoms are useful models for bulk matter while having properties accessible to direct calculation or computer simulation. They have optical, electronic and thermal properties very different from those of bulk materials of the same chemical composition.

In these small groups of atoms (in some contexts called 'quantum dots' or 'nanocrystals') most of the atoms are on the surface. Surface atoms have a higher energy than atoms within the bulk of

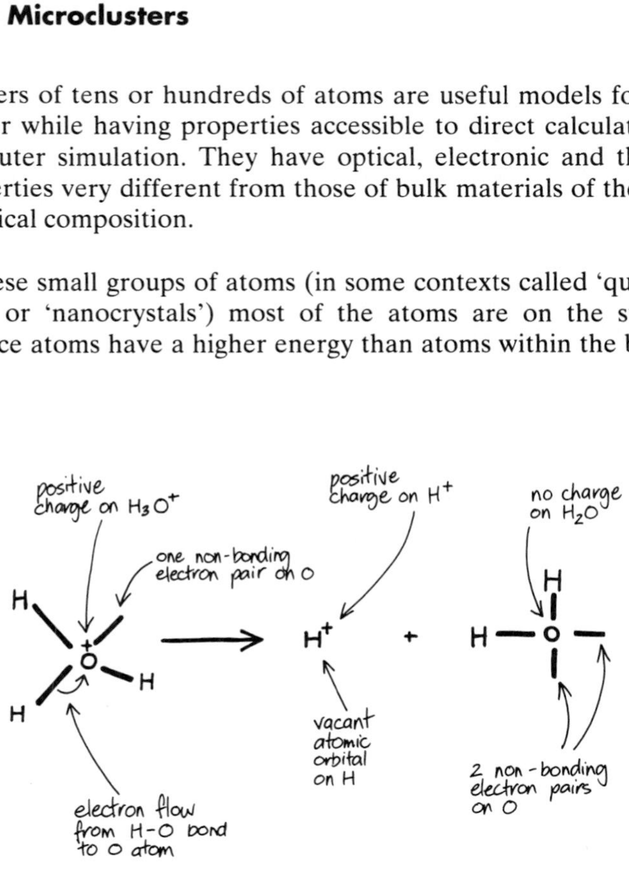

Figure 9.3 The mechanism of the ionisation reaction.

a material so in this situation the microcluster melts at a temperature well below that expected of a bulk crystal[20].

From this point of view, fullerenes are microclusters of carbon atoms, illustrating that our view of a material can change according to the model we are applying[21].

9.11 Nonexistent Compounds

W E Dasent[22] popularised the term 'nonexistent compound' with the intention of encouraging chemists to explore the reasons for

some potential compounds to be 'missing', to have low stability. More importantly from our perspective of misleading models, chemists should try not to assume that just because they have not found particular compounds, they cannot exist. In some situations the thermodynamic model 'explains' why particular compounds are unlikely to be stable, but in other cases even this information is incomplete or imprecise. There are examples of previously 'missing' compounds that have been 'discovered' almost accidentally, most notable being the compounds formed between the halogens and the noble gases, and recently the fullerenes.

9.12 Handedness

An important property of most objects in nature (such as our left and right hands) is that they are not identical with (not able to be superimposed on) their mirror images. When they have this property of 'handedness' we say they are 'chiral' (pronounced 'kyral'). This name is from the Greek *cheir* meaning 'hand', with the same origin as the word 'chiropractic' for 'manipulation' or adjustment (of the body) by hand. To distinguish the two forms, we describe them as 'right-handed' or 'left-handed', although the significance of these terms is clear only for some familiar chiral objects like hands and screw-threads[23].

Processes as well as objects exhibit chirality. Chirality extends all the way from fundamental particles and nuclear interactions, through molecules, to life-size objects like crystals, plants and animals. (We do not know whether the evolved chiral biological characteristics are a direct result of atomic and molecular chirality[24].) We can understand and use models more readily if they are intuitively attractive. 'While this does not mean that chemists think sugar molecules wear little mittens in the winter, the immediacy of "handedness" makes it a very useful tool in nomenclature and stereochemical classification.'[25]

One of the simplest chiral molecules is lactic acid. This has a central carbon atom with four 'fingers' arranged around it, all different:

- a 'methyl' group, $-CH_3$

- a 'hydroxyl' group, $-OH$

- a 'carboxyl' group, $-COOH$

- and a hydrogen atom, H.

Just as our four different fingers are related differently in our two mirror-imaged hands, there are two different ways of making a lactic acid molecule. The substituted central carbon atom we call an 'asymmetric' carbon atom, and this tetrahedral geometry of substituents in carbon compounds lends itself to the phenomenon of chirality. It is of current interest that those fullerenes described as 'carbon nanotubes' are also chiral[26].

The chemical properties of the two chiral forms of a molecule are the same, but they react differently to electromagnetic fields. In particular, they rotate the plane of polarised light in different directions, and are therefore 'optically active', or 'optical isomers'[27]. In some situations, there are equal numbers of both chiral isomers so the bulk or macroscopic mixture (described as 'racemic') is not optically active.

Crystals show up the mirror-image molecular properties on the macroscopic scale. Both mirror-image types of some compounds will crystallise from a single mixed solution as distinct crystals. Louis Pasteur in the middle of the nineteenth century observed that a salt of tartaric acid sometimes formed in wine had crystals of two types, one the mirror-image of the other. He was able to separate the crystals manually and, after making solutions of each, observed that they rotated polarised light in opposite directions[28].

Distinction between these chiral forms becomes particularly important in biological materials. As a simple example, we can differentiate the two forms of limonene because one smells like oranges and the other like lemons. More fundamentally, the chemistry of biological materials has a preferred chirality: L-amino acids and D-sugars[29]. When the molecules are immobilised in biological structures such as enzymes or in cells the differences become dramatic, with one form 'fitting' the structure better than the other. Enzymes accommodate molecules for chemical transformations in biological systems like keys fitting a lock: only the correct one works[30]. While one molecular 'enantiomer' (non-superimposable mirror-imaged compound) may be a valuable therapeutic drug, its mirror-imaged twin contributes to side effects and may even have specific harmful effects. (The Greek *enantio* means 'opposite'.) Thus it appears that only D-thalidomide causes birth defects. Most drugs from natural sources are chiral (optically pure), and an increasing proportion of synthetic drugs are now available in an optically pure form[31].

We can superimpose a stationary or even spinning sphere on its mirror image, and it is therefore not chiral (it is 'achiral'). However, if the sphere is also moving along the spin axis it becomes chiral. Our model of the electron as a moving, spinning particle is therefore chiral. Experimentally, beta particles from radioactive nuclei have chiral asymmetry, left-handed electrons far outnumbering right-handed ones. Similar asymmetry occurs in the other elementary particles. This is reflected in differences in molecular enantiomers with regard to physical properties[32].

In the Standard Model of fundamental particles the electroweak interaction distinguishes between 'left-handed' and 'right-handed' by means of 'weak charged forces' ('W forces') and 'weak neutral forces' ('Z forces'). In contrast to electric 'charges' where the force between any two electrons is repulsive, the weak W 'charge' is nonzero for a left-handed electron but zero for a right-handed electron, so a right-handed electron does not experience a W force. Therefore, nuclear beta decay, controlled by the W force, produces mostly left-handed electrons.

We gain nothing by asking **why** this asymmetry should occur. The model is merely describing our observations of the nature of matter in terms we can appreciate, if not fully understand.

Left-handed and right-handed electrons have weak Z 'charges' of opposite signs and approximately equal magnitudes. Therefore right-handed electrons are attracted to the nucleus and left-handed ones are repelled, although the effects are of extremely low strength. One result of the electroweak interactions is that the electron orbit, which would be otherwise be circular, becomes a right-handed helix in the vicinity of the nucleus. (There do not exist in Nature electrons with left-handed helical flow.) Another result is that one chiral molecule exists in a higher-energy or lower-energy state than that of its enantiomer. The differences are infinitesimal (calculations show that the number of L-amino acids should exceed D-enantiomers by one part in 10^{17}). There is debate whether this effect of the weak nuclear interaction is connected in any way with the dominance of particular chiral molecules in biological systems[33].

9.13 Aromatic Model

Cyclic molecules such as benzene (C_6H_6) have structures allowing additional electrons from pi bonds[34] to be shared in an extended chain or ring formation between several atoms. These compounds exhibit unusual structural stability. The structure based on a ring of six carbon atoms (the 'benzene ring') is particularly stable.

Chemists describe these compounds as 'aromatic' because some of the first compounds of this type studied had a fragrant odour, although some 'aromatic' compounds have no smell. Their chemical stability still tends to be discussed as deviations from the 'expected' behaviour of nonaromatic molecules. For historical reasons chemists usually describe them with a model of different structures with alternating single and double bonds that 'resonate', and the bonding electrons are 'delocalised'. They calculate the relative stability of each aromatic structure in terms of the resonance energy per delocalised electron. The resonance energy is a measure of the extra stability of the cyclic structure compared with the equivalent straight chain structure. Most stable on this basis is benzene itself[35].

The terms 'delocalisation' and 'resonance' imply that 'localisation' and separate, identifiable 'bonds' between pairs of atoms are the norm. However, this view is just the result of our preoccupation with the chemical bond model and it is preferable not to think of aromatic compounds as being at all unusual. In terms of our 'thermodynamic' model, this type of arrangement is simply the one that minimises the energy in the system. The molecular orbital model assumes that electrons are free to move throughout the whole molecule and occupy orbitals analogous to those in atoms. This model has proved particularly useful for describing and predicting the properties of aromatic compounds[36].

9.14 Fullerenes

We can achieve the 'ultimate' in aromatic compounds by the 'mind experiment' of fusing the aromatic benzene C_6H_6 rings, eliminating the hydrogen atoms, and producing a three-dimensional hollow array of pure carbon atoms in hexagons and pentagons. So far chemists have found these 'three-dimensional' carbon molecules in sooty flames and in the hot plasma products of laser-reacted graphite. In these circumstances there is enough space and time during their formation for what would otherwise be graphite 'sheets' to form themselves into three-dimensional structures. Robert Curl and Richard Smalley of Rice University and Harold Kroto (University of Sussex) won the 1996 Nobel Prize in Chemistry for their discovery of these ball-shaped forms of elemental carbon.

These hollow spheres and closed one-dimensional cylindrical tubes (nanotubes) are near-perfect crystal structures which are amongst the strongest of materials. Fullerenes are the ultimate in carbon fibres and the newest of the carbon allotropes[37]. The nanometre-diameter carbon tubes are very fine, electrically conducting wires with potentially practical applications. They have the advantage of being self-assembling materials, manufactured by processes in which fine structure organises itself without specific human direction[38]. Synthetic methods suitable for undergraduate laboratories are now available[39]. There are many other possible carbon particles, including the 'carbon onion', whose structure can be visualised readily from its name. Hard, elastic thin films of carbon can be formed by depositing high-

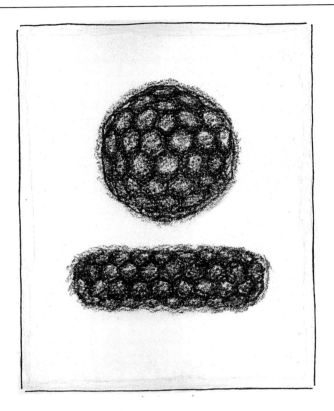

Figure 9.4 Two C_{180} fullerene molecules, one a 'ball' and the other a 'capsule'.

velocity jets of carbon onions or nanotubes on a substrate, the particles breaking and joining on impact[40].

Buckminsterfullerene itself, the almost-spherical C_{60} molecule, is most easily visualised as a soccer ball, but the white and black segments of the soccer ball model distract attention from its extremely high symmetry ('icosahedral' symmetry). Together with the fact that it undergoes rapid rotation at room temperature, this symmetry makes it a highly spherical particle. We can readily make paper models clearly showing the shapes of buckminsterfullerene and other fullerenes, and the use of paper models was important in the early studies on the structure of fullerenes. Beaton[41] has provided patterns for paper models of all 51 possible isomers for molecules from C_{60} to C_{84}. Fullerene molecules are illustrated in figures 1.3 and 9.4.

Fullerenes were synthesised from graphite in laboratory processes in connection with an investigation exploring interstellar compounds. Their first isolation from natural sources was in the solid state, in carbon-rich Precambrian rock[42]. This identification arose only because mineralogist Semeon Tsipursky was alert to the similarity of his electron microscope images to those he had previously seen of laboratory synthesised C_{60}. This was an example of serendipity in research and an illustration of the need to pursue breadth as well as depth in science. (Fullerenes apparently have not yet been detected in the interstellar medium, although evidence for other molecules has been obtained[43].)

Quantum mechanical models have estimated the relative energies of fullerenes (and of other aromatic compounds). They suggest that long fullerene capsules are higher in energy (less stable) than spherical fullerenes with the same number of atoms[44]. This is intuitively satisfying to chemists, who have learned to associate stability with symmetry. The great stability of C_{60} molecules means that physicists can use them as molecule-size projectiles in linear accelerators[45].

Fullerenes also provide a means to trap a single 'free' atom (even an atom as small as helium) within a sealed capsule made of carbon nuclei and spherically delocalised valence electrons. This isolates an atom more securely than has ever been possible before. As well as this kind of 'doping' within the fullerene molecule it is also possible to incorporate metal atoms in fullerene crystal structures. Chemists are now busy making fullerene hydrocarbons such as $C_{60}H_2$ and other derivatives based on reactions characteristic of alkenes[46].

The stimulation of research prompted by the new model of three-dimensional delocalised electron structures has already resulted in the synthesis of another class of related molecules. Metallocarbohedrenes or 'met-cars' contain twelve carbon atoms and eight titanium atoms, and other similar molecules are expected.

9.15 Macromolecules

'Macromolecule' is a rather imprecise term denoting a very large molecule, particularly applied to polymers, formed when molecules are fused together in a head-to-tail fashion by a 'polymerisation' process. Polyethylene (known as 'polythene') has repeat segments of ethylene (also known as ethene) linked together. Usually macromolecules occur in mixtures with a distribution of chain lengths or molecular masses. Some control is exercised during polymerisation to obtain the desired range of sizes. Macromolecules containing only a few repeat segments are 'oligomers'. Other terms used are 'monomer' (the unpolymerised precursor of the polymer), 'dimer' (containing two segments) and 'trimer' (three segments).

The chemistry of the polymerisation process consists of the appropriate chemical reactions, repeated many times, but with each reaction not being fundamentally different from any other organic reaction. What is important is the sequential nature of these processes. If monomers are linked head-to-tail, long and randomly coiled molecules are formed, rather like cooked spaghetti. The individual polymer molecules can be of any length, from nanoscopic to macroscopic. If we can 'bridge' or 'crosslink' polymer molecules at various places by forming bonds between reactive groups on different chains, we form inflexible three-dimensional structures, again of various lengths.

Another approach is to devise a system in which the polymer molecules can develop in a predictable way: self-assembling materials like the nanoparticle fullerenes. One way of doing this is to rely on 'fractal polymerisation'. We can use the way in which a tree grows as a model of this process. The single stem of a small tree forms branchlets, which in turn become branches that develop more branchlets, and so on, the branching often being seasonally controlled. Now at the molecular scale, imagine a core molecule with four reactive sites on to which we attach four monomer molecules, each with growing sites of their own. In this situation the branching will increase exponentially; for example, there will be 4, 8, 16, 32, ... branches successively as the process continues and the molecule grows. We can control the chemical characteristics of both the interior of this polymer molecule and its outer surface. Thus we can provide the outermost surface with many functional groups[47].

One area of increasing interest is the organisation of repeated processes of 'replication' or molecular copying[48]. Methods have been developed for synthesising

- 'isotactic' polymers (all 'asymmetric'[49] carbon atoms having the same configuration), and

- 'syndiotactic' polymers (with alternating configurations), rather than

- 'atactic' polymers (arising from completely random events).

If methyl methacrylate (the monomer) dissolved in dimethylformamide polymerises in the presence of isotactic poly(methyl methacrylate), the rate of polymerisation increases and syndiotactic poly(methyl methacrylate) is formed[50]. Some of the properties of macromolecules are determined by their particular functional groups (alcohol, ketone, etc), and in this respect they are no different from ordinary organic molecules. Other properties, however, are strongly influenced by the long-chain nature of these macromolecules.

Enzymes are a very special type of macromolecule. They are naturally occurring structures with the function of facilitating chemical reactions by positioning reagents correctly. Enzyme molecules have active sites or chemical units separated by the right number of amino acid 'spacer' units to provide just enough flexibility to adjust to the correct conformation. (The right number of units turns out to be thirteen or fourteen.) Enzymes, of course, provide our best examples of chemical catalysis and of replication[51].

Endnotes

[1]One mole of a substance contains as many molecules (or atoms if the material is made up of nonassociated atoms) as there are atoms in exactly 12 grams of carbon-12, that is, approximately 6.022×10^{23} particles.

[2]Further reading Gorin (1994).

[3]Further reading Kolb (1977).

[4]Further reading Norrby (1991).

[5]The corresponding quantity for bulk solids is called the 'work function'. Electrons are more difficult to remove from single atoms than from bulk materials: most metals have ionisation energies about twice the value of their work functions. One of the interests in the studies of microclusters (**9.10 Microclusters**) is the way the ionisation energy changes as the number of atoms in a cluster increases.

[6]Further reading Davies (1980, Chapter 3).

[7]For example Rouvray (1994). See also **2.6 Limits and Dimensions**.

[8]Nelson (1994).

[9]Fortman (1994).

[10]See **13.2 Crystals**.

[11]The forces are van der Waals forces: see **10.1 Molecular Interactions**. The properties of liquid carbon, experimentally inaccessible, have been predicted by computer simulation (**10.4 Computer Simulation**) as metallic with an electrical conductivity similar to that of the liquid metal mercury. Further reading Gillan (1993).

[12]Further reading Hammond and Kuck (1992), Curl and Smalley (1991), Huffman (1991), Maddox (1993).

[13]Fullerenes can be considered not only as elementary forms of carbon, but also as aromatic compounds which are so arranged that they contain only carbon: see **9.13 Aromatic Model** and **9.14 Fullerenes**, and also as microclusters (**9.10 Microclusters**) with their special properties.

[14]See **5.9 Dynamic Models**. Any graphite formed during the deposition is etched away by atomic hydrogen. Further reading Geis and Angus (1992).

[15]Further reading Selinger (1989).

[16]See **1.2 The Greeks Had a Word for It**.

[17]Further reading Harrison (1993), Nelson (1990, 1993).

[18]Further reading Hoffmann (1993). Roald Hoffmann asks the question 'Chemists can create natural molecules by unnatural means. Or they can make beautiful structures never seen before. Which should be their grail?'

[19]Wentland (1994).

[20]See also **7.7 Quantum Dots** and **14.2 Interfacial Energy**. Further reading Duncan and Rouvray (1989), Goldstein *et al* (1992), Berry (1990), Corcoran (1990), Reed (1993).

[21]Further reading Curl and Smalley (1991).

[22]Further reading Dasent (1963, 1965).

[23]Further reading Heilbronner and Dunitz (1993, Chapter 7).

[24]Further reading Hegstrom and Kondepudi (1990), Avetisov *et al* (1991).

[25]*Journal of Chemical Education* (1994). Further reading Barta and Stille (1994), Thall (1996).

[26]Further reading Dresselhaus (1992).

[27]See **4.2 Wave States**. The mirror-image forms are also known as 'enantiomers': L-enantiomers and D-enantiomers, L- and D- standing for *laevo-* (left) and *dextro-* (right) from early optical rotation studies.

[28]The nucleation (**13.3 Nucleation**) of a supersaturated solution by the first-formed crystal can result in the separation of many crystals with the same handedness, an example of replication (**15.9 Replication**). Further reading Orgel (1992).

[29]Further reading Amato (1992c), Hegstrom and Kondepudi (1990), Orgel (1992), Avetisov *et al* (1991).

[30]See **15.10 Biological Matter**. Synthesis of macromolecules (**9.15 Macromolecules**) with controlled optical configurations is also possible. Further reading Orgel (1992).

[31]Further reading Thall (1996).

[32]Diagram: Hegstrom and Kondepudi (1990).

[33]Further reading Hegstrom and Kondepudi (1990), Avetisov *et al* (1991). Paul Davies (1980, Chapter 5) has explored a model of space being closed but with a 'twist' in the nature of a Mobius strip that would result in the chirality of an object being reversed on 'circumnavigation'.

[34]See **9.9 Carbon Compounds**.

[35]Further reading Aihara (1992).

[36]The ability of bonding electrons in aromatic compounds to interact with matter and radiation in other ways makes them of considerable interest. In particular, there are the fullerenes (**9.14 Fullerenes**), aromatic compounds containing only carbon (**13.5 Organic Conductors**) like polyaniline and the van der Waals molecules (**10.2 Fragile Molecules**) that include aromatic molecules hydrogen-bonded to water. Further reading Heilbronner and Dunitz (1993, Chapter 10).

[37]See **9.6 Allotropes**. Further reading Hammond and Kuck (1992, p ix), Curl and Smalley (1991), Curl (1992), Huffman (1991), Guo *et al* (1991), Dresselhaus (1992), Ebbesen and Ajayan (1992), Ajayan and Iijima (1992), *Journal of Chemical Education* (1992), Ge and Sattler (1993), Heilbronner and Dunitz (1993, p 62), Maddox (1993), Taylor and Walton (1993), Ball (1996).

[38]Further reading Whitesides (1995).

[39]Further reading Iacoe *et al* (1992), Craig *et al* (1992).

[40]Amaratunga *et al* (1996).

[41]Beaton (1992, 1995).

[42]Further reading Amato (1992a), Buseck *et al* (1992), Emsley (1993).

[43]Further reading Arnau *et al* (1995).

[44]Further reading Adams *et al* (1992), Maddox (1993).

[45]Further reading Aldous (1992).

[46]Further reading Smalley (1991), Hammond and Kuck (1992, p ix), Emsley (1993), Taylor and Walton (1993), Hebard (1992).

[47]See **3.8 Fractals**. Further reading Tomalia (1995). See also **15.9 Replication**.

[48]See **15.9 Replication**.

[49]See **9.12 Handedness**.

[50]Further reading Orgel (1992). See also **15.2 Polymers**.

[51]Further reading de Gennes (1992), Orgel (1992). See also **12.9 Catalysts and Enzymes** and **15.9 Replication**.

CHAPTER 10

MOLECULAR MATTER

Once we can visualise matter in the form of atoms or molecules, we can explore the ways of combining molecular models to develop models for bulk materials.

10.1 Molecular Interactions

This section deals with the physical interactions between discrete atoms or molecules. Physical interactions are those that do not result in the formation or breaking of chemical bonds. In this model, electrons remain in the vicinity of their own molecular particle (atom or molecule), rather than being more widely dispersed.

(Here I use the term 'molecular particle' to mean both 'atom' and 'molecule' and the word 'molecular interaction' to include both 'molecular interaction' and 'atomic interaction'. This is because the noble gases at most temperatures and many other chemical elements at high temperatures exist naturally in the atomic rather than molecular form. The properties of gases made up of isolated atoms and those comprising small molecules show similarities.)

The interactions between molecular particles (attractive or repulsive depending on circumstances) chemists and physicists call 'van der Waals interactions' after the nineteenth-century physicist Johannes van der Waals. The thermodynamic model tells us that the equilibrium position reached in the arrangement of molecular particles in a system is governed by the balance between

- lowest total energy, and

- highest total entropy or disorder. Also,

- the influence of entropy decreases with decreasing temperature[1]†.

Some molecules are polar, that is, they have some regions with a permanent net negative charge and other regions with a net positive charge. There are interactions between these polar molecules that we can visualise in a classical electrical model. The negative region of one molecule attracts the positive pole of a neighbouring molecule, and vice versa (a 'dipole–dipole attraction'). In addition, each permanent dipole induces a temporary dipole ('induced dipole') in adjacent molecules, resulting in a dipole–induced dipole attraction. The extent of interactions of this type strongly influences the melting and boiling points of polar substances. Other things being equal, a polar substance has a higher melting point and a higher boiling point than a nonpolar substance.

We can visualise another type of interaction between atoms or molecules (whether polar or not) with the electron cloud model[2]. This considers an electrically negative cloud of electrons surrounding the positively charged nucleus (in an atom) or group of nuclei (in a molecule). We consider the electron cloud of even a nonpolar molecular particle (one having no permanent dipole moment) as becoming temporarily displaced so that it is no longer centred on the nuclei. This results in one part of the molecular particle having a net excess of positive charge and another part an excess of negative charge: fluctuating dipole–induced dipole interactions. On balance, when molecular particles are close there are net attractive interactions ('dispersion interactions' or 'London interactions') between neighbouring molecules.

There is a more fundamental model useful in describing these

† Endnotes for this chapter can be found on page 232.

interactions. Molecular quantum electrodynamics (QED) leads us to another way of looking at molecule–molecule interactions as well as the interactions between molecules and light. The 'zero-point radiation field' which always accompanies matter in this model can account for an interaction between molecules any distance apart. One molecule in its ground state acquires energy temporarily from the radiation field, allowing it to achieve an excited state and emit a 'virtual' photon. Another molecule subsequently absorbs the photon, which is then re-emitted, and finally re-acquired by the original molecule. Both molecules are back in their ground states and the net energy is unchanged. This process can be thought of as going on all the time between all molecules to varying extents[3].

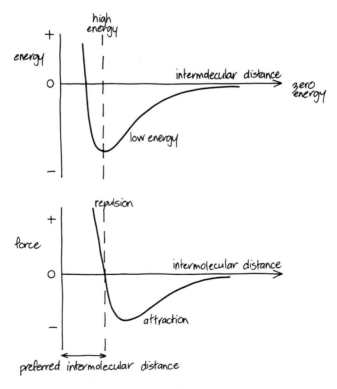

Figure 10.1 Energy and force models for intermolecular processes.

The result is that as molecular particles approach each other there is a net attractive interaction that first increases (up to a certain proximity). Eventually, as the particles become even closer, the electron clouds begin to overlap and the attractive interaction passes through a maximum, decreases, and finally becomes strongly repulsive (figure 10.1). Molecules therefore cannot get too close together. In other words (now using the energy model) the molecular particles reach an intermolecular distance that minimises the total energy. Although there is some associated ordering, this effect of decreasing entropy is usually insignificant compared with the energy advantage. Repulsive interactions are of much shorter range than the attractive interactions, which are typically effective over several molecular diameters[4].

In molecular crystals, molecules pack together to minimise 'empty space'. We can interpret this as attractive and repulsive interactions being balanced (energy minimised): a convex surface of one molecule tending to be next to a concave surface of another, and so on[5].

Terms such as 'molecular interactions' usually refer to

● cohesive interactions between atoms or molecules in homogeneous gas, liquid or solid phases (including intimate mixtures like solutions), and

● adhesive interactions between dissimilar molecular particles in adsorption situations[6].

However, in appropriate circumstances the same interactions are important between **macroscopic** objects[7].

Molecules containing atoms of high electronegativity (like oxygen) can strongly attract the bonding electron cloud away from a hydrogen atom, leaving it with a small positive charge. This allows the (now slightly positive) hydrogen atom and the (now slightly negative) oxygen atom of another molecule to attract each other, forming a 'hydrogen bond'. It is as if two oxygen atoms on different molecules share the hydrogen atom, which forms a chemical bridge between them. This weak bond is intermediate in strength and lifetime between the (rather weak, nonspecific) van der Waals molecular interactions on the one hand and the (relatively strong) chemical bonds on the other. (The distinction between physical and chemical interactions was illustrated in figure 5.2.) Liquids with hydrogen-bonding molecules (like those with polar molecular interactions) tend to have high melting points and high boiling points. The best known example of hydrogen bonding is in water, H_2O. The hydrogen-bonding model accounts for some of the unusual properties of water. One of these is that ice is less dense than liquid water, whereas the solid forms of most materials are denser than their liquid forms.

Hydrogen bonds occur only when the molecules have the approximately correct orientation. Because they require orientational ordering, there is loss of orientational entropy when hydrogen bonds form. This parallels the loss of positional entropy when a disordered liquid crystallises into an ordered solid. Molecules which are not hydrogen bonded are free to take on any orientation and therefore they have a higher orientational entropy. This entropy penalty in hydrogen-bonded systems becomes more significant at higher temperatures, so hydrogen bonds are more stable at lower temperatures. In a dimer comprising two water molecules at low temperature in the vapour phase, one unit is the hydrogen bond donor and the other is the acceptor.

Hydrogen bonding is significant in interactions between organic molecules and water. For example, the acetylene molecule (C_2H_2) donates hydrogen to an oxygen of water. In the case of benzene (C_6H_6), the aromatic pi electron cloud accepts a hydrogen bond from **both** hydrogen atoms of a water molecule positioned close to it. These are examples of 'van der Waals molecules'[8].

10.2 Fragile Molecules

There are particular specific molecular interactions which are rather stronger than most van der Waals interactions (although not as strong as chemical bonds that hold the atoms within 'normal' molecules). These cause extensive short-range and short-term ordering described as 'weak bonding'.

The thermodynamic model accounts for these weakly bound molecules held together by van der Waals interactions existing in preference to the unbound particles. In the gas phase, atoms of noble gases or molecules of hydrogen and nitrogen can associate in this way. At very high pressures there is evidence for a solid van der Waals compound between helium and molecular nitrogen, $He(N_2)_{11}$. It is likely that at the pressures existing in planetary interiors, compounds of this type are common (as long as the components are sufficiently abundant).

We can model the stability of solid van der Waals compounds with packing densities, using 'hard spheres' for the component atoms of noble gases or simple molecules such as nitrogen, N_2. We observe experimentally that mixtures of two sizes of macroscopic spherical particles with diameters having ratios of 1.00 to about 0.85 arrange themselves randomly. However, if the diameters are very different (with ratios less than about 0.57) the smaller particles can occupy the spaces that occur between the larger particles in their normal packing arrangements. If this happens, the larger particles determine the packing density and the smaller ones have little effect. In both cases (mixtures with very similar diameter ratios and with very different diameter ratios) the compositions of the mixtures are continuously variable and there is no tendency for compound formation.

Applying these findings to a mixture of helium and nitrogen, we find that the ratio of the diameters of the nitrogen molecule to the helium atom is between 0.6 and 0.7. This is intermediate between the quoted limits. The van der Waals compound, under the driving force of a tendency towards high-density close-packing at high pressures, is denser and therefore more stable than an unbound array of separate particles. This is true although the atoms interact only weakly in 'chemical bond' terms[9].

Extensive cooperative van der Waals interactions of the weak bonding type occur in condensed phases (liquids and solids) in the appropriate circumstances. However, we may study them best in the gas phase as isolated van der Waals molecules. We achieve this by carrying out the studies under rather unusual conditions, such as in supersonic jets of material. Here the conditions are equivalent to those in an ordinary gas at ten kelvin (minus 263 degrees Celsius, ten degrees above absolute zero) and the compounds are in their ground states (lowest vibrational levels)[10].

10.3 Many-Body Problems

We use the term 'many-body problem' to describe any assembly of more than two objects. The prediction of the behaviour of even small numbers of objects as a function of time is surprisingly difficult. The application of newtonian mechanics to mechanical and astronomical problems over the past three hundred years has involved analytical solutions to equations of motion (difficult for even three objects) and numerical approximation techniques[11]. There is also the choice between 'atomistic' or particle mechanics and 'continuum' mechanics depending on the circumstances. The latter is far more effective and appropriate for engineering problems and the former is necessary for simulation of materials at the molecular level. Computer simulation uses both techniques.

10.4 Computer Simulation

Since the 1920s, theoretical chemists and physicists have had a major goal: to predict the properties of materials from first principles. The best model for this appears to be the quantum mechanical wave equation describing the energy and distribution of electrons in atoms and molecules. Much earlier than this the applied mathematicians of the eighteenth and nineteenth centuries who followed Isaac Newton developed analytical methods for the solution of problems in classical interactions of 'continuous' matter. These methods now bear their names: Leonhard Euler, Joseph-Louis Lagrange, Karl Gauss and William Rowan Hamilton. Many of the particular goals they pursued still elude us. However, the computing abilities we now have do enable us to carry out simulation of the behaviour of groups of atoms in a molecule or of assemblies of molecules. This is the process of 'molecular dynamics'. The images generated provide an additional and very effective way of generating models of systems which are otherwise inaccessible to us. An example is a time-lapse view of the way a molecule interacts with a solid surface or of the probability of finding a particle near another particle in a fluid. We can convert spectroscopy results into molecular likenesses that we can more readily understand. We must be aware, however, that built into these models are the approximations and assumptions made elsewhere in our models. If we animate the interaction of ball-and-stick molecular models we may obtain valuable information, but it will still be a poor approximation to reality. If we simulate a liquid made up of molecules with a particular type of mathematically described intermolecular potential we are relying on the realism of the chosen mathematical model[12]. The properties of microclusters are especially appropriate for computer calculation because of the limited number of atoms involved[13].

10.5 Ideal Gases

The simplest description of a gas is probably 'a substance that fills any container it occupies'. We can model this as composed of hard, elastic, fast-moving particles which repeatedly collide with each other and with the walls of the container. An important quantity in the thermodynamic model is internal energy. When the internal energy of a material at constant pressure is large, the molecular particles (molecules or atoms) of which it is composed travel relatively long distances between collisions. As a result, the shape and size of the container rather than the particular nature of the molecular particles control the motion of the particles (and the 'shape' of the material).

The mean free path is the average distance which the particles travel between consecutive collisions, so distances of this order are important in determining the properties of gases[14]. The corresponding time interval (the mean free time) represents a typical time scale for the gas.

To make the study of gaseous systems simpler we often model a perfect gas or ideal gas as

- having particles with mass and velocity (and therefore energy),

- occupying negligible volume, and

- not interacting with each other or with their container, except for

- perfectly elastic collisions with each other and their container.

The physical properties of the common real gases (including nitrogen and oxygen in air) approximate to those expected of the ideal gas model reasonably well. For this reason, the results of some early physical chemistry experiments on the properties of gases conducted in the seventeenth century were particularly easy to interpret. We still name the various 'gas laws' describing these properties after the early scientists:

- Robert Boyle ('the volume of a gas at constant temperature is inversely proportional to the pressure')

- Jacques Charles and Joseph Gay-Lussac ('the volume of a sample of gas at constant pressure increases linearly with increasing temperature').

It is interesting that the technological requirements of balloon flights stimulated these investigations. For the model of gases as being composed of invisibly small and well separated particles in random motion we are indebted to early Greek science. The idea was revived by Daniel Bernouilli in 1738 and again by James Clerk Maxwell over a hundred years later. Amedeo Avogadro in 1811 made the derived proposal that the volume of a gas depends only on the number of particles it contains. In modern terms, 'equal volumes of all gases at the same temperature and pressure contain the same numbers of molecular particles'. The mathematical relationship or ideal gas 'law' linking pressure, volume, temperature and chemical amount that results from combining the equations of Boyle, Charles and Avogadro is the equation of state for ideal gases. This turns out to be a very simple relation[15]. It is the starting point for more realistic (and mathematically complex) equations of state that provide better approximations of the properties of real materials.

Because all low-pressure gases have very similar properties, the properties of mixed gases follow directly from the composition. We can treat each component of a mixed gas as if it occupies the container alone, a result described by John Dalton in the early nineteenth century. The behaviour of gases when we mix them is far simpler than we observe for the mixing of any other combination of states of matter.

The early investigations, simple observations and successful modelling of gas properties were of great importance in developing our understanding of matter. This work generated the model based on molecular particles which were very small compared with the extent of their influence. It meant that chemists had a good estimate of the actual number of molecules in a given amount of gas. They could select a known number of molecules simply by taking a particular volume of gas at a particular temperature and pressure.

10.6 Hard Sphere Model

In gases the molecular particles interact in rather particular ways.

- The repulsive interactions between atoms or molecules are only short range.

- The attractive interactions between nonpolar particles are often rather insignificant compared to their kinetic energies.

Because of this combination of properties, we can model some materials as if they had the properties of macroscopic 'hard spheres'. We characterise hard sphere interactions by the absence of remote effects until the actual collision, and abrupt elastic repulsion at the point of impact. This model works reasonably well for the larger atoms like argon, and for approximately spherical molecules. We can describe the interactions mathematically with simplicity, facilitating both calculations and computer simulation.

10.7 Molecular Speeds

The kinetic molecular model of ideal gases requires that in gas particles the average kinetic energy (which varies directly as the square of the velocity) is proportional to the absolute temperature (expressed in kelvin). Just as important as the average energy is the way that the particles distribute energy between themselves. Even if we could start off with all particles having the same velocity, repeated collisions would randomise the distribution. The energy of a particular particle depends on the nature of its last collision, even at constant temperature, with some travelling much faster than others. Because of the large number of particles and the large number of collisions, we can predict with great

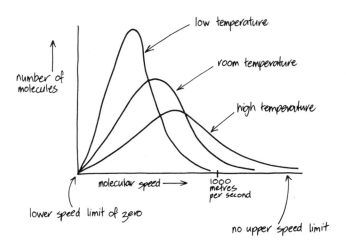

Figure10.2 Maxwell distribution of molecular speeds.

accuracy the fraction of the molecular particles having any narrow range of energies.

At room temperature, the molecules in the air around us are moving at average speeds of about five hundred metres per second (eighteen hundred kilometres per hour). Most individual molecules have speeds somewhere between zero and a thousand metres per second. (Molecules of hydrogen move three to four times faster than this, and atoms of mercury vapour have mean velocities of only about two hundred metres per second.) The distribution curve (figure 10.2) is not quite symmetrical. The curve has a zero limit on the left-hand side, and there is a high-energy 'tail' because there is a small chance of a few particles having extremely high energies. The detailed shape depends on the temperature, with the tail on the right-hand side being markedly greater at higher temperatures. The average must increase at the higher temperature, and at the low-energy end the speed cannot be less than zero, so the proportion of high-speed particles must increase. A very small proportion of molecules have speeds close to zero and another small proportion have particularly high speeds, but most have speeds close to the average value. Both theoretical (model) and experimental results agree that the distribution of speeds (the distribution curve) is a peak rising from zero to a maximum and then falling less sharply than it rises. The curve is a graph of the number of particles with particular speeds plotted against their speeds (or energy), and the area under the curve represents the total number of particles.

This is a graphical model of the energy distribution situation in the gas. If you have any difficulty accepting this you can think of the same kind of curve representing everyday information. If you plotted the annual income of a group of people in this way, you would get a similar curve. There would be a small number with zero income, a small number with very high income, and most earn around the average wage. It was first worked out by James Clerk Maxwell towards the end of the nineteenth century, and we often call it the 'Maxwell distribution' of speeds.

Higher-mass molecules have lower average speeds at a given

temperature, and the distribution of speeds is narrower than for lighter molecules.

10.8 Kinetic Fluid Model

Although our perception of gases as being composed of molecules in random motion has a long history, scientists have modelled gases in different ways. One of the earlier models tried to answer the question of what keeps the molecules apart by proposing that each molecule was surrounded by a shell of 'ether' that rotated at high speed. In this model it was the centrifugal force exerting the interaction with neighbours that kept the molecules apart. The resulting predicted properties were very similar to those derived in our current ideal gas model, as one would expect from any model describing gas pressure with newtonian kinetic energy ideas.

Real gases have properties which deviate from the ideal gas model because atoms or molecules do in fact interact with each other at a distance to some extent. At low pressures and large volumes the molecules are so far apart most of the time that the molecular interactions are insignificant, and the gas is close to ideal. At rather higher pressures and smaller volumes (that is, higher densities) such that the molecules are separated by distances of the order of molecular diameters, the attractive interactions become important. At still higher pressures we have to include the repulsive interactions in the model and we can no longer regard the molecular size as zero. In a dilute gas only 'binary collisions' (collisions involving just two molecular particles on any one occasion) are significant. In a denser fluid this is no longer true and the system is much more complex. Much of the progress has been the result of computer simulation on a limited number of particles[16].

Endnotes

[1]Although van der Waals forces do not usually result in the formation of recognisable 'chemical bonds' there are special circumstances where weakly bound molecules of this type can be considered to be formed: see **10.2 Fragile Molecules.** Further reading Walker and Vause (1987).

[2]Described in **7.4 Electron in a Box.**

[3]See **4.4 Heisenberg May Have Been Here!.** Further reading Craig (1992), Feynman (1985).

[4]Discussed in many books and articles, such as Barton (1973, 1974) and Jennings and Morris (1974).

[5]Molecular crystals are discussed in Chapter 13. Further reading Fagan and Ward (1992).

[6]See **14.15 Adhesives.**

[7]See **14.11 Sticky Solids.**

[8]See **9.13 Aromatic Model.** In terms of **5.1 Energy Well Model,** the potential well of the benzene aromatic ring has a relatively 'flat' bottom, making it an excellent hydrogen-bond acceptor. Further reading Klemperer (1992), Suzuki et al (1992) (including cover illustration of computer simulation of the van der Waals molecule). For water, see Amato (1992b) for example.

[9]See also **13.8 Solid Solutions** and **5.6 Cussedness Model.** Further reading Vos et al (1992).

[10]Further reading Klemperer (1992). See also **15.1 Soft Matter.**

[11]See also **3.3 Newton's Model.** Further reading Hoover (1984).

[12]Further reading Amato (1992e), Hoover (1984), Evans et al (1984), Suzuki et al (1992), Gillan (1993), Orszag and Zabusky (1993).

[13]See **9.10 Microclusters.** Computer simulation is particularly useful for studying processes resulting from random events such as the formation of aggregates (**14.13 Aggregates**) and systems at criticality (**2.4 Condition Critical**). Simulation is now extending beyond single materials and providing information on chemical reaction kinetics (**12.8 Chemical Processes**). In the field of nanotribology (**14.14 Nanotribology**), the time scales (many nanoseconds) and distance scales (tens of nanometres) of experimental investigations and computer simulations of physical surface interactions can now be overlapped so that the reliability of simulations can be assessed. However, when considering the growth of crystal grains it is unlikely that it will ever be feasible to start with atoms as units. Further reading Belak (1993a,b), Landman and Luedtke (1993), Robbins et al (1993), Harrison et al (1993), Finnis (1993).

[14]See **2.3 How Long Is a Piece of String?.**

[15]See also **12.4 Phase Diagrams.** For an amount n of an ideal gas, and any pressure

Image not available

(p), volume (V) and temperature (T), $pV = nRT$, where R is the gas constant, 8.314 J K^{-1} mol^{-1}.

[16]In particular, correlations between successive collisions have been found to be more significant than previously anticipated. Also, at high shear rates some aspects of the behaviour described in **15.4 Flowing Polymers** are exhibited. Further reading Cohen (1984).

CHAPTER 11

DISORDERED MATTER

Liquids play an essential role in human activities and in life itself, but the liquid state of matter leads a rather precarious existence. Each covalent compound is stable as a liquid over a relatively narrow range of physical conditions, although metals and salts tend to be liquid over wider ranges.

11.1 Condensed States

The quantum model, so successful in describing the structures and properties of atoms and of molecules, also accounts for the existence of condensed materials.

The common characteristic of condensed phases is their space-filling ability. If one calculates the space available for each molecular particle (atom or molecule) as if it were free and gaseous, then the particles would have to be in contact. The most successful models of condensed phases also take into account the sharing of electrons. Electrons that one would otherwise consider to be the 'outer' electrons of individual atoms or molecules are not completely localised in this way. Instead, all molecules throughout the material tend to 'share' the outer electrons, but to different extents according to the nature of the material.

Materials exist in condensed phases as liquids and solids over certain ranges of temperature and pressure because in the appropriate circumstances these states are more stable than the gaseous one. The thermodynamic model provides energetic advantages in having molecular particles arranged in close-packed configurations, either random or ordered. The strong attractive or cohesive interactions existing between the particles cause considerable negative internal energies compared to vapour phase molecular particles. (Vapours have negligible amounts of potential energy of this kind.) The internal energy is negative because one has to supply energy to a condensed system to convert it to a gas or vapour. If the particles are ions, even stronger attractive interactions arise from coulombic electrostatic interactions in materials such as metals, ionic crystals and ionic liquids.

The model of 'cohesion' has proved to be very useful. The 'cohesive energy' is equal to the internal energy but with the opposite sign, so it is positive rather than negative. It provides cohesion and holds the molecular particles together in a condensed phase. In the right circumstances the cohesive interactions even in liquids can withstand considerable external tension[1]†.

The two extreme models of ordering in condensed materials are:

- Crystalline solids, with perfect, infinite, periodic or ordered array, unchanging with time, and

- Liquids, with no long-range order, with short-range order fluctuating with time, but on average appearing the same in all directions ('isotropic').

- There are intermediate types of condensed matter, such as 'colloidal crystals' and 'liquid crystals'[2].

Any 'order' in the molecular particles making up materials may be of two kinds:

- positional (the only type possible with spherically symmetric particles), and

† Endnotes for this chapter can be found on page 253.

- orientational (particularly apparent with long, rod-like molecules).

11.2 Liquids

A liquid has a fixed volume but (when in a gravitational field) has a shape determined by the lower part of its container. Over the (relatively narrow) temperature range for which each material exhibits these properties the internal energy is such that there is considerable freedom of motion, with short-range cohesive and repulsive interactions reasonably balanced. Liquids exist only at temperatures where the thermal energy is of the same order as the energy of interaction between molecules. Once the thermal energy increases significantly beyond that the liquid evaporates, and if the thermal energy decreases much below that the liquid solidifies to a crystal or in a metastable state to a glass.

The liquid state is much more difficult to visualise or describe than either the highly ordered crystalline state or the completely random gaseous state. Models of liquids may start with the regularity of a crystal then introduce random particle-sized 'holes' to produce 'fluidity' (ability to flow) and disorder. Alternatively, they combine aspects of crystal models and gas models to achieve the same result. A liquid has some short-range order or structure simply as a result of the molecular particles being packed together fairly closely, but none of the long-range order that exists in a crystal. The nature of the short-range order is also time dependent. To some extent liquids are 'solid-like' at short times and short distances but 'fluid-like' at long times and long distances[3].

Water is an integral part of our existence and people tend to think of it as a 'typical' material. Unconsciously, they make water our 'model' for the behaviour of simple molecules. This is far from the truth, and the reason that water is such a useful biological material has much to do with its unusual and atypical properties. Hydrogen bonding plays an important part in these properties. Physical chemists and chemical physicists have developed numerous models to describe water structure, often proposing that water molecules combine through their hydrogen bonds into short-lived two-dimensional or three-dimensional arrays[4].

11.3 Superfluid Helium

A 'superfluid' is a fluid that can flow without 'friction', the two isotopes of helium being the known examples.

As one cools helium gas at atmospheric pressure, it remains liquid at even the lowest temperatures. Consistent with Heisenberg's uncertainty principle and the very light and weakly interacting nature of helium atoms with their large zero-point motion, solid helium does not form without considerable external pressure. Helium condenses to a liquid at 4.2 kelvin. If one continues the cooling, at a temperature near 2 kelvin (minus 271 degrees Celsius) helium acquires very different properties. It flows without viscosity or friction, it can pass through gaps too small for gases to escape and it has a very high thermal conductivity. The origins of superfluidity (the frictionless flow of liquid) are quantum mechanical in nature, although the particular situations for the two isotopes of helium are rather different[5].

Helium-4 is a boson, indicating that it conforms to Bose–Einstein statistics and that that any number of particles can occupy the same quantum state simultaneously. At temperatures close to absolute zero (below the 'lambda point', 2.18 kelvin) almost all atoms are in the lowest energy state. Considerable energy (more than can be supplied by most normal collisions) is necessary to raise any atom to a higher energy state. Some of the random excitations are sufficiently energetic to do so, and these excitations transmit through the superfluid, 'colliding' with each other and with the container walls.

Physicists model the behaviour of superfluid helium above absolute zero as a mixture of two interpenetrating components, a superfluid at absolute zero and a 'normal' fluid carrying any energy present. The proportions of each form change with temperature, from 100% normal at the lambda point to 100% superfluid at absolute zero. The model considers the superfluid component to be without viscosity and carrying no entropy. The normal fluid component does have viscosity and entropy. It is possible to cause the superfluid fraction to flow without friction in a closed loop packed with fine emery powder, which prevents the normal fluid from moving. This model helps us to understand the very high thermal conductivity of superfluid helium. The 'normal' atoms carry the energy down the temperature gradient while the superfluid flows in the opposite direction without carrying energy. The two fluids slip past each other without friction, a process very much easier and faster than diffusion.

From a quantum mechanical point of view helium-3 is fundamentally different from helium-4. Its atoms contain an odd number of neutrons and therefore an odd number of particles overall and so are fermions, unable to condense into the same ground state. Consequently, helium-3 cannot become superfluid as easily as helium-4. However, at very low temperatures (millikelvin temperatures, about a thousandth of a degree above absolute zero) a weak pairing of atoms occurs. The result is that the angular momentum of each pair is exactly the same, so they are particles of the boson type[6] that can condense to a common ground state and exhibit superfluidity. The pairs have spin properties, and because spin belongs to the liquid as a whole, superfluid helium-3 (unlike helium-4) has directional properties (rather like those of liquid crystals). This provides a 'texture' that can be aligned by surface forces such as magnetic fields, surfaces or liquid flow.

The mechanical behaviour of superfluids differs from our views of what is 'normal' for a liquid. If we rotate a container of liquid helium, the liquid does not rotate as a whole as we might expect. The normal fluid moves with the container, but not the superfluid. All the atoms in the superfluid, being described by the same quantum mechanical wave function, must have the same momentum. This is not possible if the liquid rotates as a whole because it would require the momentum to be proportional to distance from the rotation axis. Instead of uniform rotation of the whole superfluid sample, there is 'flux quantisation'. Miniature quantised 'vortex' lines develop within the liquid. These run parallel to the rotation axis of the container and form a pattern, the 'vortex lattice', which physicists can observe experimentally. The dimensions depend on the rotation speed. ('Vortex' means 'whirlpool'. Physicists use a comparable idea for magnetic flux in superconductors.) The size of the vortex core is about a tenth of a nanometre, less than an interatomic distance. It is a node in the quantum mechanical wave function describing the superfluid. A convenient combination of the particle and wave models of behaviour allows us to say that the wavelength of each atom depends on its circular velocity within the vortex. Because a whole number of wavelengths must fit into the circular path, the rate of movement is quantised. The 'normal' fluid also present in this superfluidity model does not circulate around vortex cores, but interacts with them in other ways.

The vortices cause frictional losses at the walls, a departure from superfluidity. Vortices form most easily when the flow is unrestrained, with the result that superfluids flow more easily through fine pores than through coarse ones. Scientists associate 'quantum turbulence' or 'superfluid turbulence' with the unpredictable bending and twisting of vortex cores. They also use this as a 'model' system for the study of classical turbulence in ordinary liquids, because in some ways it is easier to 'understand' or to model than classical turbulence[7].

11.4 Viscous Flow

For the 'ideal' case of a nonviscous (inviscid) fluid without internal friction, flow would continue even without a potential gradient. This is what occurs in superfluid helium.

For some situations even in 'ordinary' everyday fluids it is useful to adopt this simple model of nonviscous fluid flow. We gain insight into fluid behaviour, particularly of the simpler, low-viscosity fluids, from a model which does not dissipate kinetic energy as thermal energy. In this case, in applying the energy conservation model we consider only kinetic energy, pressure–volume energy and gravitational energy. We neglect frictional losses and elastic energy.

This simple model predicts that the pressure in a flowing fluid is lower than it is in a stationary fluid. You can see this in several practical situations:

- In the 'atomiser', high-speed air passing across the top of an open vertical tube dipping in a liquid causes a decrease in pressure, so the liquid rises up the tube.

- A spinning ball moves in a curved path, because the velocity of the surface on one side of the ball relative to the air is greater than that on the other.

- On a railway platform there is a risk of being 'pushed' by atmospheric pressure into the low-pressure region created by a fast-moving train.

- The upthrust on the wing of an aircraft occurs because the air velocity over the wing is greater than that under it.

That flowing 'real' fluids have some frictional conversion of energy to heat does not affect the general nature of these properties. They are just as applicable to real, viscous fluids as they are to ideal, nonviscous fluids.

A viscous liquid needs a potential gradient (a pressure) for flow. There are actually two flows occurring. The most obvious one is the volume flow or mass flow of the fluid itself (parallel to the streamlines in the case of laminar flow) as a result of the pressure gradient. The transport coefficient describing the ease of flow is the 'fluidity', which is the reciprocal of the 'dynamic viscosity coefficient'[8].

However, in viscous flow there is also a momentum flow, normal (perpendicular) to the streamlines. (We can compare this with heat conduction from hotter to colder regions.) During viscous fluid flow there is a transfer of momentum from the faster flowing regions to the slower flowing regions. The density of momentum flux is proportional to the velocity gradient, the proportionality constant or medium factor being the 'dynamic viscosity coefficient'.

The dynamic viscosity coefficient, which we often call just 'viscosity coefficient' or 'viscosity', is a characteristic of the material, describing the amount of internal friction resisting its flow. The greater the viscosity, the smaller is its mass flow rate or volume flow rate (flux) in a given situation. Conversely, the greater the viscosity the greater is the momentum flow within the liquid **perpendicular** to the flow.

A common situation involving viscosity occurs when an object (relatively rapidly) acquires a terminal velocity or limiting velocity while falling under gravity through a viscous fluid. The fluid may have a high viscosity (like treacle) or a low viscosity (like air, where the falling object can be a parachute). Similar effects provide the drag on a rising cavity ('bubble') in a liquid[9].

The field of science dealing with flows on the macroscopic scale is 'hydrodynamics' or 'fluid dynamics'. It uses conservation of mass, momentum and energy together with empirical relationships linking fluxes, potential gradients and proportionality coefficients to develop fluid flow models. There are certain limitations. One is that the system must be large compared to the mean free path (or the time scale must be longer than the mean free time, as we saw in the case of ideal gases). Another is that the gradients should not be so steep that the relations defining the transport coefficients become nonlinear. Until recently we thought that the model used in these hydrodynamic relations was inappropriate for objects as small as molecules. However, the 'generalised hydrodynamics' model has extended hydrodynamics to the molecular scale. Chemical physicists do this by allowing the transport coefficients to depend on the distances and times for which the potential gradients act. (Fluctuations within fluids can occur over a wide range of time scales.)

The subject of fluid dynamics dates from the seventeenth century when Isaac Newton defined the viscosity coefficient relating the shear stress to the velocity gradient. This simplifying equation and those derived from it perform well under conditions of

- 'ordinary' time scales (seconds or minutes), for

- 'ordinary' macroscopic samples of fluids (often known as 'newtonian fluids' or 'simple liquids'), made up of

- 'ordinary' molecules (with relative molecular masses below about one thousand).

For these liquids, along a horizontal pipe of uniform cross-section the pressure drop is proportional to the flow rate, as observed by Poiseuille. Solutions and melts of macromolecules do not conform to this behaviour and we describe them as 'non-newtonian' or 'viscoelastic'. They have properties that depend on both the viscous and the elastic characteristics of the materials. (It is interesting that Poiseuille was an anatomist, concerned with blood flow in the body. Although his study led to a relationship between flow rate, tube diameter and pressure which is appropriate for newtonian fluids, blood is non-newtonian.) All forms of 'soft matter' (Chapter 15), from colloidal systems to rubbery materials, have significant elastic properties as well as viscous properties.

The viscosity coefficient as a measure of internal friction or intermolecular resistance to flow bears some resemblance to the elastic moduli describing intermolecular resistance to elastic disruption[10]. The difference is that

- the elastic moduli relate the applied stress to the resulting strain, while

- the viscosity coefficient relates the applied stress to the strain **rate** of the response.

The viscosity coefficient is the flow equivalent of the elastic rigidity. As long as the viscosity coefficient is constant as we change the stress, we say the flowing material is 'newtonian', the flow analogy of Hooke's law for elastic deformation.

Materials not showing this behaviour are 'non-newtonian', and are typically gels and pastes containing large (polymeric) molecules that interact only weakly with each other[11]. Even with small shear stresses the structures in these gels tend to break

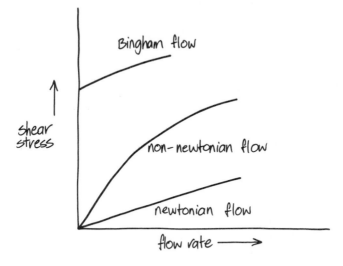

Figure 11.1 Typical flow processes in terms of flow rate as a function of shear stress.

down to produce a more 'newtonian' material, resulting in non-linear relationships between stress and flow rate (figure 11.1). An example of these 'thixotropic' materials is water-based gel paint. This paint conveniently flows like a liquid when we stir it or brush it, but 'stays put' like a gel without flowing down the wall when we remove the shear stress. Materials like clays exhibit another kind of behaviour, 'Bingham flow'. These materials can withstand a fairly high initial shear stress without flowing. However, once clay begins to flow, water released from between the microcrystalline plates assists the flow process. This property allows clay pottery to be shaped easily but to retain its shape before and during firing.

However, things are not simple even for 'simple' fluids. A dense, hard-sphere material, when subjected to a shearing force, distorts so that its packing is less dense and its total volume increases, the phenomenon of 'shear dilatancy'[12].

The time-dependent behaviour of material is clearly important, just as the time scales of interactions are critical in dynamic models of any kind. According to the prophet Deborah 'the mountains flow before the Lord'[13]. This expression emphasises the importance of time in discussing solids and liquids. Mountains flow on the Lord's time scale but not on ours. There is even a 'Deborah number', defined as the ratio between the characteristic time for the process under study, and the length of the observation time. If the observation time is of a human scale, solids have large Deborah numbers and liquids have small numbers. However, a very short observation time can effectively increase the Deborah number.

There are three natural frequencies or time scales important in flow processes:

- The duration of a collision between molecules. This is instantaneous for 'hard sphere' model materials[14], but is generally of the order of a femtosecond (a millionth of a nanosecond)[15] for real materials.

- The time between collisions, the mean free time. This is important in the kinetic model of fluids, and is typically tens or hundreds of femtoseconds.

- The time scale for the validity of the hydrodynamic model. We once though this to be much longer than the first two, but we now know it to be of the order of a hundred femtoseconds[16].

What we are doing is generalising the hydrodynamic model so that it applies at molecular time scales and molecular distance scales. This amounts to including viscoelastic effects in our model, even for 'simple liquids' made up of small molecules such as water.

Whether a particular material acts as a flowing 'liquid' or as a nonflowing and rebounding 'elastic' depends on the time scale of the force applied. The silicone polymer Silly Putty® illustrates this. It deforms and flows as a viscous liquid if compressed slowly, but bounces like rubber if dropped on the floor.

- In everyday life we are more aware of the **viscous properties** of the relatively free-flowing materials like water. These exhibit elastic properties only over such short time scales and such short distance scales that the elastic effects are usually not noticeable.

- We usually observe the **elastic properties** of very viscous materials like plastics and glasses which exhibit flowing properties only over unobservably long times[17], or of crystalline materials that fail by brittle fracture rather than flow[18].

- However, all materials demonstrate both viscous and elastic properties to some extent: they are all **viscoelastic**.

Each material at a particular temperature has a characteristic time, the 'relaxation time', separating 'short' times when elastic properties dominate from 'long' times when viscous properties dominate[19]. For observation times of the order of the relaxation time we must consider the full 'viscoelastic' properties. Otherwise we can use a 'viscous' model alone for experimental times longer than the relaxation time, or an 'elastic' model alone for very short times. Although all materials are viscoelastic, if possible we use simpler models that consider either the viscosity or the elasticity alone, rather than both. In practice, the simpler models are inappropriate and the full viscoelastic model must be used to describe the behaviour of

- 'simple' liquids in short-duration or high-frequency situations,

- glasses over very long times, and

- complex liquids (soft matter) on any time scale, as explained later[20].

All this relates to the bulk movements of molecules in homogeneous fluids[21]. The mobility of materials in the fluid phase influences the rates of chemical processes, and also sets the maximum limit on the overall rate. However, for non-homogeneous systems there is another important model.

If one material is moving through another and is dispersed within it as molecular particles the process is 'diffusion'. The flow-rate or flux is proportional to the concentration gradient (Fick's 'law'), and the proportionality coefficient is the 'diffusion coefficient'. The diffusion rates of reactants or products and not chemical reaction rates control the overall rates of many processes. These are 'transport-controlled reactions'. Even if there is bulk movement as a result of thermal convection or stirring, there are relatively undisturbed liquid layers next to reaction sites. Through these 'boundary layers' molecular diffusion is the limiting transport process.'

11.5 Glasses

Our everyday use of the word 'glass' refers to those metal silicates that provide us with 'window glass' and a wide range of specialty glasses. These include the heavy metal compounds like lead silicate with a high refractive index ('lead crystal', which is definitely a glassy material and not a crystalline solid). We require both the thermodynamic (equilibrium) model and a dynamic model[22] in discussing the properties of glasses. A glass is a solid-like material made by cooling a liquid at such a rate that it is 'locked in' to its disordered or random structure. It is slow to undergo rearrangement from this metastable state to the equilibrium or thermodynamically stable crystalline state. (Even slow cooling is sufficient for very viscous materials like liquid sugar, tars or metal silicates to achieve a glassy state.) All glasses have a tendency to crystallise or devitrify, and this process involves shrinkage because ordered packing is more efficient than disordered packing. This causes some of the older glass artefacts to have weakened, although it is possible to make special glasses which devitrify in a reasonably short time and are useful as glass ceramics.

A glass can serve as a static model of the structure of a liquid at any instant, as long as we make allowance for the increase in density that occurs during glass formation. Like solids, glasses have high Deborah numbers so they do not flow when we observe them (on our human time scale). However, over a longer period (as in the glass windows of very old cathedrals, for example) the flow is sometimes measurable. Conversely, of course, 'ordinary liquids' are 'very viscous' on short time scales. The structural and

physical properties of glasses are very different from those of crystals. Scientists are now using the concept of 'medium-range order' for the characterisation of 'amorphous' (glassy) solids[23].

There is a very narrow temperature range over which the viscosity changes dramatically from one that we would describe as a 'very viscous liquid' to one that we would call a 'glass'. This is the 'glass transition temperature'. It is not an 'ordinary' phase transition in thermodynamic terms, and it is not as sharp as a melting point, but the temperature range is sufficiently short for it to be useful in the characterisation of glasses.

The ultralow-temperature properties (such as sound velocities) of glasses differ from those of crystalline materials and are being studied down to the microkelvin temperature range[24].

We can describe the glassy state as 'frozen chaos': a combination of rigidity and chaos. Surprisingly, recently physicists have modelled the liquid–glass transition by means of a stack of two mutually perpendicular sets of parallel superconducting wires with a tunnel junction at each intersection. Glass transition processes, being slow, are difficult to observe or simulate, whereas superconducting arrays have analogous transitions that can be readily studied electronically[25].

A glass containing a small proportion of vanadium oxide and very small (nanometre scale) pores can act as a colour-change indicator for the presence of various gases. Gases include water vapour, ammonia, formaldehyde and hydrogen sulphide. The gas molecules diffuse into the pores, react strongly with the vanadium oxide, modify the way light interacts with vanadium, and so cause a colour change. These 'xerogels' with vanadium or other metals are therefore potential indicators for a range of gaseous pollutants in the air[26].

11.6 Turbulence

'Turbulent flow' replaces 'laminar flow' when certain conditions of criticality like flow velocity and pipe narrowness are exceeded.

In helium superfluidity we saw that fluid turbulence is so complex that we use quantum vortex behaviour as a model for it. In turbulent systems there is disorder on all scales. There are temporary structures that dissipate energy, with large movements and eddies being degraded to smaller ones, right down to the molecular level. At the same time, small irregularities can grow to large sizes. Sometimes turbulence suits us, as when we are trying to mix things. Sometimes it does not, as when we are trying to maintain streamlined air flow over the wings of aircraft to keep them flying. On one side of a set of critical conditions small disturbances die out, but on the other they grow dramatically so they can affect the whole material. The laminar plume from a heat source (like the smoke rising gently from a burning cigarette tip) continues until the flow exceeds a critical rate and turbulent eddies develop.

It is believed that turbulent vortices nucleate at or near the walls containing a flowing fluid. Trace amounts of polymer molecules on the walls apparently reduce turbulent drag by inhibiting this nucleation. In terms of an energy model, the 'elastic' energy stored in the macromolecule can exceed its kinetic energy, providing a capacity to influence the flow process. Although scientists do not understand this process well, the Alaskan oil pipeline uses the effect[27].

'Convective turbulence' arises when a fluid is heated from below. The hotter (less dense) regions tend to rise and the colder, denser regions to fall, and if the flow velocities are high the motion is irregular in space and time. However, it is not entirely random: there are characteristic patterns and structures. Most obvious are 'thermal plumes', the mushroom-shaped dissipative structures made familiar by nuclear explosions. Thermal plumes are columns of heated material rising from a small heat source at the bottom of a fluid and expanding outwards at the top of the fluid. In the case of nuclear explosions, the atmosphere is the fluid and the nuclear explosion is the 'small' heat source. Thermal plumes are also associated with volcanic activity and atmospheric 'thermals'.

Another structure exhibited in the process of convective turbulence (and also in other dynamic processes) is the 'boundary layer'. This is the undisturbed hot layer of fluid at the surface of the heat source, or the concentrated solution at the surface of a reacting or dissolving solid. It is a layer maintained at a steady state thickness by a feedback process.

In the thermal boundary layer example, if there is a tendency for the layer to become thicker than a certain value, it becomes unstable and multiple plumes rise. If it is becoming thinner, fewer plumes rise and thermal conduction causes the heated layer to thicken. The boundary layer thickness depends on the depth of the fluid because it is influenced by the full plume structure. As with all systems resulting in a steady state, our perception of the structure and process differs according to our point of view. It changes according to whether we are looking at the boundary layer, the thermal plumes or the whole system[28]:

- the boundary layer thickness is predictable,

- thermal conduction into the boundary layer is predictable,

- an **isolated** plume has a predictable structure,

- thermal transfer through the system as a whole (neglecting the details of the boundary layer and plumes) is predictable, yet

- the detailed structure of plumes varies unpredictably in space and time.

11.7 Granular Matter

Assemblies of solid particles have some properties reminiscent of solids and others of liquids, and are best described as being in a 'granular state'[29]. This state of matter is highly dissipative[30], and supports temporary structures when provided with an external energy source. A vibrated granular medium has properties analogous to those of viscous liquids. In this case, the macroscopic patterns are composed of 'oscillons': localised, particle-like structures. A layer of small grains with oscillations vertically driven from below forms oscillons which are of two phases (a 'peak' or a 'crater' at a particular time). Oscillons of like phase repel each other, while unlike oscillons attract. As a result, geometrical patterns form, rather like those of crystals composed of positive and negative ions.

We can 'fluidise' a bed of granular matter by means of an upward flow of air or other gas introduced from beneath it. Energy is being provided from outside (by the air flow), maintaining the granular material in a fluid-like structure. The material in the fluidised bed acts in some ways like a liquid, with an effective density that depends on the air flow rate. Fluidised beds can thus separate materials on the basis of density differences. Fluidised beds of fuels are also incorporated in furnaces. Here the coals themselves are the granular materials being fluidised by the air or oxygen that is also required for combustion.

We notice that piles of sand or similar granular materials have a critical slope, the 'angle of repose'. (A pile of dry sand makes an angle of about 34 degrees with the ground.) One grain added to the top of a pile with sides steeper than the critical value can fall and disturb others locally, triggering an 'avalanche' process. If, on the other hand, the sand pile has not yet achieved this critical angle, the steepening of the pile continues. The system seems to search for a slope that is 'just right'. We use the word 'criticality' in a particular sense in science for processes where events of all sizes can occur[31]. The process can also be modelled by computers[32]. Thus granular heaps do not flow unless the angle of repose is exceeded (in contrast to liquids, which flow at even the smallest angle), and when flow occurs, it is only in the surface layer. (Unlike the fluidised bed, where all the granular material is in a fluid state provided with energy from the air stream, in this case it is only the surface layer that is 'fluidised', the structure being supplied with gravitational energy.) If a snow-covered mountain slope is steeper than the angle of repose for snow, snow avalanches can occur.

Beneath a liquid, the pressure depends on the depth of liquid, regardless of how deep it is. However, if a tall container is filled with granular material beyond a certain level, the pressure at the bottom is **independent** of the depth. The frictional effects of grains with the walls and with each other provide support. An extension of this behaviour is the pressure under a pile of grains arranged in the shape of a pyramid or solid cone. (Imagine an industrial pile of bulk material, or even a large pile of fruit on display for sale.) We might expect that the maximum pressure would be at the central point of the base, directly under the apex or highest point. In fact, the pressure measured there can be 30% less than it is elsewhere on the base. Computer solution of 'classical' mechanical equations[33] for the frictional motion of each particle in a reasonably large pile is still 'too hard' to reproduce these experimental results[34]. The behaviour of sound in granular materials is also unexpected: it is not transmitted in the simple manner of a homogeneous material but depends critically on the contacts between individual particles.

The densest possible packing of (uniform spherical) particles in three dimensions is 'close-packed hexagonal' (filling 74% of the space)[35]. Computer simulation shows that the equivalent 'random close-packed' limit is a volume fraction or density of 64%. As the packing is made more tenuous, eventually the loosest possible random packing is reached (only one continuous path of grain–grain contacts throughout the array) at a density of 56%, the 'random loose-packed' limit. It follows that the variation in average inter-grain distance from the dense, random close-packed limit to the tenuous, random loose-packed limit is only about 10%. Only a small change in effective density is necessary to convert a compact granular material into a fluid-like state.

Granular systems as simple as random close-packed arrays of hard spheres show 'shear dilatancy' (decreased density of packing under shear). A close-packed granular material can only deform or flow if expansion occurs. Osborne Reynolds demonstrated this in 1885. He packed a leather bag with marbles, filled it with water, and saw the water level drop when he twisted the bag. The water had occupied the extra space created between the less closely packed marbles. You can observe this effect yourself when walking on wet sand at the beach: the area around your foot 'dries out' . The (shear) deformation of the structure resulting from the localised force exerted by your foot is accompanied by expansion, with the result that there is more free space and the water level drops[36].

Endnotes

[1]See also **5.1 Energy Well Model**, **5.9 Dynamic Models** and **6.1 Interactions**. It turns out that cohesive energy density (cohesive energy per unit volume) is particularly useful, giving rise to cohesion parameters (**12.6 Cohesion Parameters**) as quantities that can be used for prediction (**12.5 Mixtures** and **15.3 Polymer Solvents**). Cohesion parameters are sometimes known as 'solubility parameters' or 'Hildebrand parameters'. The dimensions of cohesion parameter are square root of pressure: when the cohesion parameters for two materials are multiplied together the cohesive pressure (or cohesive energy density) is obtained. Further reading Barton (1975, 1983a, b, 1987, 1990, 1991).

[2]See **15.6 Colloids** and **15.8 Liquid Crystals**.

[3]Further reading Barton (1973, 1974).

[4]The process of nucleation (**13.3 Nucleation**) is necessary for the crystallisation of ice from water and also for vaporisation (**5.9 Dynamic Models**). Further reading Amato (1992b).

[5]See also **4.4 Heisenberg May Have Been Here!**, **4.5 Playing Cards and Four-Strokes**, **4.6 Quantum Leap**, **3.5 Flow** and **11.4 Viscous Flow**.

[6]These are known as Cooper pairs: see **4.5 Playing Cards and Four-Strokes** and **13.7 Superconductors**.

[7]Further reading Bardeen (1990), Donnelly (1988, 1995), Lounasmaa and Pickett (1990), Pobell (1993), Tilley and Tilley (1990), Bernstein (1996). Not all atoms in superfluid helium-3 are paired: there are excitations associated with each unpaired atom and the empty state of its shadow particle or 'hole': see also Chapter 13.

[8]Further reading Kaye (1993, Section 6.3). See also **3.5 Flow**, **3.7 Chaos** and **11.3 Superfluid Helium**.

[9]Further reading Shafer and Zare (1991).

[10]See **6.5 Failure**.

[11]See also **15.4 Flowing Polymers**.

[12]See **11.7 Granular Matter**.

[13]Judges 5:5. Further reading Evans *et al* (1984).

[14]See **10.6 Hard Sphere Model**.

[15]10^{-15} second.

[16]Further reading Hanley (1984), Alder and Alley (1984).

[17]Further reading de Gennes (1984).

[18]See **6.5 Failure**.

[19]See **2.2 Perspective**. These effects may be discussed also in terms of a

characteristic frequency, being the reciprocal of the relaxation time.

[20]See **15.4 Flowing Polymers**.

[21]Above, and **3.5 Flow**.

[22]See **5.9 Dynamic Models**.

[23]Further reading Elliott (1991).

[24]Further reading Pobell (1993).

[25]See **3.7 Chaos** and **7.7 Quantum Dots**. Further reading Wolynes (1996).

[26]Geake (1994).

[27]Further reading Witten (1990).

[28]Turbulence was referred to in **3.7 Chaos** and in **5.7 Ordering**. See also **2.4 Condition Critical** and **3.2 Dynamics**. Further reading Kadanoff (1991), Mandelbrot (1983), Ruelle (1991), Gleick (1988).

[29]Further reading Jaeger and Nagel (1992), Fineberg (1996).

[30]See **5.7 Ordering**.

[31]See also **2.4 Condition Critical**.

[32]See also **10.4 Computer Simulation**.

[33]See **3.3 Newton's Model**.

[34]Further reading Watson (1991).

[35]See **13.2 Crystals**.

[36]There is further discussion in **10.6 Hard Sphere Model**, **11.4 Viscous Flow** and **15.4 Flowing Polymers**. Further reading Evans *et al* (1984), Jaeger and Nagel (1992).

CHAPTER 12

MIXED MATTER

For most practical purposes in our daily activities we use mixtures of materials rather than pure compounds. Solutions (mixtures with a liquid as at least one of the components) are particularly important in environmental and biological systems as well as in industrial processes.

12.1 Purity

As I have already emphasised in several contexts, it is important to understand the particular significance of words used in the description of scientific systems and their models. The apparently simple terms 'pure substance', 'mixture' and 'solution' often cause problems for new students. We all think we know what the word 'mixture' means, but even as used by chemists this term is rather ambiguous. One meaning is that it is a blend at an unspecified level of subdivision of two or more pure substances not linked by chemical bonds: not a compound. You can separate a 'mixture' of iron filings and sulphur powder with the aid of a magnet, but not the iron and sulphur in the compound iron sulfide. (A pure substance, when recovered from a mixture, retains its definite composition and specific properties. There are specific linkages between the atomic units of matter that we can most readily describe or model as 'chemical bonds'.)

Another meaning (the one I shall use here) is much stricter. In a mixture of two or more components the blend must be 'homogeneous' (uniform in composition—otherwise it is 'heterogeneous') right down to the molecular level, with a completely random distribution of the components. This is the meaning assumed when mixing is referred to as an equilibrium state in the thermodynamic model. When one of the components is a fluid (liquid or gas) we can achieve such a homogeneous mixture reasonably readily, but a thermodynamically defined mixture of two solid components is much more elusive.

Chemists apply the term 'solution' to homogeneous mixtures in which there is more of one component (the 'solvent') than the other (the 'solute'). If the interactions between solute and solvent are particularly strong, we say that the solute is 'solvated'. Solvents and solutes may be gases, liquids or solids without restriction. When a polymer is in equilibrium with a liquid there must be two phases present. The liquid phase is a dilute solution of the polymer in the liquid and the polymeric phase is a dilute solution of the liquid in the polymer. Just as it is possible to have a metastable situation in a pure material that we 'supercool' to below its normal freezing point, it is possible also for a solution to become 'supersaturated'. In other words, the solvent contains more solute than it would if the system were at equilibrium[1]†.

A word whose meaning is being modified by popular use is 'alloy'. Originally, it referred to a mixture of metals that ideally was homogeneous at the molecular level. (In practice, some metallic mixtures described as alloys are 'homogeneous' only at the microcrystal level, random arrays of pure crystals of different metals being visible by optical microscope or electron microscope.) In different contexts, the word has been both broadened and specialised. Polymer scientists call a homogeneous mixture of two or more polymers a 'polymer alloy'. An example of a more specific use is in 'alloy wheels', the light-weight and aesthetically pleasing car wheels made from an aluminium alloy.

† Endnotes for this chapter can be found on page 273.

12.2 Ideality

In an attempt to portray the behaviour of homogeneous mixtures, physical chemists have devised simple model systems, analogous to ideal gases. Most homogeneous systems involve a liquid.

I would like you to visualise the open surface of a liquid in equilibrium with its vapour, a dynamic equilibrium with vapour-phase molecular particles striking the liquid surface, transferring their energy and becoming incorporated in the liquid. At the same time, the jostling of liquid particles on the surface provides some of them with enough energy to leave the liquid and become part of the vapour. When a particular pure liquid is in equilibrium with its vapour at a particular temperature, the vapour pressure of that substance has a definite value. In the ideal mixture model, the vapour pressure of each component is proportional to the amount[2] of that component present. The proportionality constant is the vapour pressure of the pure substance, so the vapour pressure of course reaches that of the pure liquid when it is present to the extent of 100%. (This is 'Raoult's law'.) If two components are present in equal amounts, each would have a vapour pressure equal to one-half that of the corresponding pure compound. In an ideal solution, then, both solvent and solute would conform to Raoult's model.

This is not a very useful model for most practical systems. In real solutions, although the vapour pressure of the solute may be proportional to its amount at low concentrations, the proportionality constant is not that of the vapour pressure of the pure substance. Therefore physical chemists use an ideal dilute solution model that conforms to this 'less ideal' behaviour, known as 'Henry's law'.

In thermodynamic terms, an ideal mixture or ideal solution is formed only when two very stringent conditions apply:

- One of these is that there is no enthalpy change on mixing (ideal or zero heat of mixing).

- The other is that the entropy change on mixing corresponds to that of a completely random process (ideal entropy of mixing).

A less restrictive model (one that is closer to reality for at least some systems) is one in which the two kinds of particle distribute themselves randomly. This means an ideal entropy of mixing despite a nonzero (nonideal) enthalpy of mixing, and is the 'regular solution' model. The behaviour of this model approximates to that of real solutions made up only of molecules which are uncharged, nonpolar and not capable of undergoing specific interactions such as hydrogen bonding.

12.3 Collections

There is another specialised word that chemists use in describing the properties of collections, because there are some properties (such as temperature and pressure) which are meaningless for single particles. All collections, animate or inanimate, exhibit properties that we cannot predict easily from a knowledge of the individuals. In the case of collections of molecular particles we describe these properties as 'colligative'. The **numbers** of particles rather than the types of particle determine the properties of gases. In the same way, some of the properties of solutions depend on the number of particles and not their actual masses. These properties are all associated with solvent vapour pressures[3]

- the effect of dissolved substances on the freezing point of the solvent

- the effect of dissolved substances on the boiling point of the solvent

- osmosis of a solvent through a semipermeable membrane in a solution.

12.4 Phase Diagrams

Our model tells us that higher-temperature phases (like water vapour) are less ordered than lower-temperature phases (like ice crystal) and this works in most situations. However, there are exceptions that warn us that this model is much too simple. The thermodynamic model treats this behaviour in terms of the balance or equilibrium resulting from the interplay between energy on the one hand and disorder or entropy on the other, and is the usual model chosen to discuss equilibrium and phase changes.

It is clear both from observation and from the thermodynamic model that pressure, volume and temperature are important variables in the properties of materials. It turns out that we can define the state[4] of a system containing only one kind of substance by specifying the values of four variables: the amount of the substance, pressure, volume and temperature. Further, these four quantities are not independent, but are linked by an 'equation of state', which is a more complex version of the equation of state for an ideal gas[5]. There are only three independent variables necessary to specify the state of each phase. If there are two phases of a pure substance present (for example, ice and water) there are only two independent variables. The location of the 'triple point' where solid, liquid and gas are present simultaneously is beyond our control: a pair of values for temperature and pressure exists that is characteristic of the

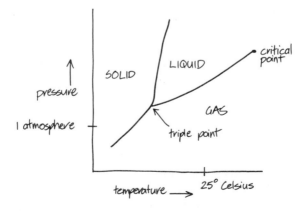

Figure 12.1 The phase diagram for carbon dioxide.

substance. Physical chemists often present this information as a two-dimensional phase diagram. A complete diagram would be three-dimensional (pressure, volume and temperature) but we can project the phase lines that occur in three dimensions onto the two-dimensional (pressure versus temperature) diagram. This is convenient, and there is little loss of information. The one point where these lines intersect (three phases existing together) is the 'triple point'. A gas is a 'vapour' at temperatures below the critical temperature, and a 'supercritical fluid' at pressures and temperatures above the critical point.

As an example of the way that phase diagrams are used, the phase diagram for carbon dioxide (figure 12.1) shows the triple-point pressure (where the gas, liquid and solid are all present at equilibrium). For carbon dioxide, the triple-point pressure is greater than atmospheric pressure, so you can immediately determine that the liquid phase of carbon dioxide cannot exist without external pressurisation.

In discussing the properties of materials we must avoid placing too much emphasis on features which are 'accidental' due to the particular atmospheric pressure and temperature range on the surface of the Earth. For example, the boiling point of water has no universal significance because it depends on the atmospheric pressure. The triple point (ice, water and water vapour in equilibrium) is a more fundamental property, because it is independent of external pressure.

Physical chemists and chemical engineers use phase diagrams not only for solids, liquids and gases, but also for all other states of matter. For example, they model the way gels swell to different extents under varying conditions of solvent and temperature[6].

Imagine a liquid and vapour in equilibrium, with both pressure and temperature increasing. The liquid becomes increasingly volatile, and the vapour becomes increasingly dense. The critical point is the particular combination of pressure and temperature where the distinction between liquid and gas disappears. Above the critical point there is only a single phase, the 'supercritical fluid'. However, when the conditions are very close to the critical point there are drops of liquid and bubbles of gas that have the potential to be of all sizes, from a single molecule up to the macroscopic dimensions of the whole sample[7].

12.5 Mixtures

When we pour two liquids into a container and stir them together they may form a single homogeneous phase: they are miscible. If, however, the liquids remain distinct as separate coexisting phases (separated by the phase boundary or 'meniscus') we say that they are immiscible. However, this really means 'not fully miscible'. It is important to realise that neither of the resulting two 'immiscible' phases is identical to the original liquid. If you vigorously mix an oily liquid and water, then allow them to separate, it will appear that the oil is floating unchanged on top of the water. However, the oily phase is not pure oil but 'wet oil', a homogeneous mixture, a very dilute solution of water molecules dispersed in oil. Likewise, the watery (or aqueous) phase is not pure water but a very dilute solution of oil molecules dispersed in water. The system reaches an equilibrium or balance that the thermodynamic model describes with energy and entropy terms for each of the two phases. In molecular terms, there is a dynamic process in which molecules of water and oil are continually moving back and forth across the phase boundary so that the concentrations on each side remain constant. In macroscopic (that is, bulk liquid) terms we can describe the state of the system by specifying two quantities at each temperature. These quantities are the solubility of the oil in water and the solubility of water in the oil, that is, the composition of each of the phases.

The temperature is a significant variable in determining the position of equilibrium. Some binary liquid systems may be either miscible or immiscible, depending on the temperature. If the two immiscible liquids are to become miscible (introducing more disorder), it is usually necessary to increase the temperature. However, there are examples of this miscibility occurring as the temperature decreases. Sometimes we even observe that as the temperature increases, two immiscible phases at first become miscible then at a higher temperature immiscible again, the

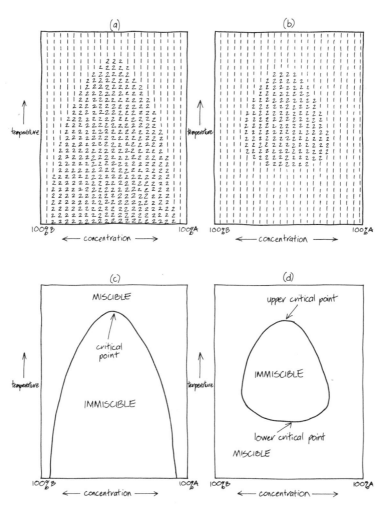

Figure12.2 Phase diagrams for liquid–liquid systems. The boundaries between the observations of one phase and two phases in (a) and (b) correspond to the lines in (c) and (d).

phenomenon of 'reappearing phases'. To understand this phenomenon more fully we can use a graphical model, a phase diagram, drawn from experimental information (mutual solubilities at each temperature).

Consider an experiment in which we mix equal amounts of two liquids, shake them up, then wait to see whether they separate into two layers or remain as a single phase. We repeat this at different temperatures and with different proportions of the two liquids, drawing a diagram by entering '1' if a single phase results and '2' if there are two phases (figure 12.2(a)). This is quite typical behaviour: miscibility often increases with increasing temperature. A second, less common type of phase diagram for mixtures of two liquids is shown in figure 12.2(b). The information is displayed more clearly in figures 12.2(c) and 12.2(d). In each case there are single-phase regions (situations of miscibility) and two-phase regions (immiscibility), but in one case the immiscible phase is 'closed' (entirely surrounded by the miscible phase). The line separating these regions is the 'coexistence curve'. These diagrams summarise a large amount of information about these liquids, telling us whether a particular mixture of the two liquids is miscible or immiscible at a particular temperature. If the point representing a particular composition and a particular temperature lies within the single-phase region, the two liquids are miscible. If not, they are immiscible. The particular combinations of composition and temperature known as critical points have special properties. A relatively small number of pairs of liquids have a coexistence curve in the form of a closed loop, (d). Such systems have two critical points.

We can appreciate more easily what is happening on a molecular scale to produce these macroscopic effects by invoking some of the models previously introduced. In particular, physical chemists use the thermodynamic model, the van der Waals interaction model and the hydrogen-bonding model. For most systems, the attractive interactions between like molecules are greater than those between unlike molecules. The energy of the system is lower if each molecule is surrounded by other molecules of the same kind. If this were the only factor most combinations of liquids would be immiscible. However, the position of equilibrium is determined not only by the minimisation of energy, but also by the maximisation of entropy or disorder. This latter effect becomes more important as the temperature rises. A homogeneous mixture has additional compositional entropy because the two types of molecule are distributed randomly rather than being collected together in each phase.

In most systems, at low temperatures the liquids are immiscible (because this situation minimises their energy). At high temperatures they are miscible (because this maximises their entropy). However, for systems in which hydrogen bonds are important the interplay of energy and entropy can cause other types of phase diagram as a result of orientational entropy.

Consider a system in which only unlike molecules can undergo hydrogen bonding. There is no hydrogen bonding in either of the pure liquids, just in the mixture. At high temperatures, the two components are miscible because the high compositional and orientational entropy effects dominate when the unlike molecules intermix randomly. At intermediate temperatures the minimisation of energy starts to become more important than the maximisation of entropy, so the van der Waals energy gains from like molecules gathering together outweigh the entropy gain effect. However, as the temperature decreases still further the energy gains from hydrogen bonding between unlike molecules in a single phase have an even greater effect, and miscibility reappears[8].

The relative strengths of hydrogen bonding in the two pure liquids and their mixture determine the particular shape of the experimentally determined phase diagram. It is useful to identify the various models used in developing our understanding of liquid–liquid immiscibility:

- a graphical model (the temperature versus composition phase diagram),

- a molecular model (hydrogen bonding),

- an interaction model (van der Waals interactions),

- a thermodynamic model (interplay of energy and entropy effects).

Each model we use provides information on one aspect of a problem, but we can combine them so that with experience they form an integrated intellectual process. Eventually we use what I

call scientific 'intuition' to predict what will happen in a given set
of circumstances without analysing exactly which models we have
used.

Time scales in fluid mixing processes can be as short as a
millisecond in small aqueous systems. They can be as long as
hundreds of millions of years, as in the mantle of the Earth.
However, they have common features. The process of mixing
involves the stretching and folding of elements (small regions) of
fluid, followed by the return of some elements to their initial
locations. These processes characterise chaotic behaviour. In most
mixing situations, although there are regions of well mixed
behaviour, there are also poorly mixed regions. There are 'islands'
of regularity which fluid can neither escape from nor enter and so
remain unmixed with the bulk of material. This chaotic behaviour
coexists with order or symmetry, and there are structures
remaining constant or evolving regularly with time. Consideration
of these effects, particularly by computer simulation, can result in
improved mixing procedures in practical systems[9].

12.6 Cohesion Parameters

'Cohesion parameters' describe the extent of cohesion within
condensed materials and of 'adhesion' between condensed phases.
In condensed phases, strong cohesive interactions exist between
the molecular particles, each with considerable negative potential
energy compared to a vapour phase particle. (Gases have a
negligible amount of potential energy originating in this way.) The
internal energy of a condensed material is its potential energy
compared to that of the ideal vapour at the same temperature.
(This energy by convention is negative in sign, so one defines the
cohesive energy as the internal energy with its sign changed,
making the cohesive energy a positive quantity.) The cohesive
pressure or cohesive energy density is the cohesive energy per
unit volume, so the volume or density of a material is also
significant in assessing its cohesive properties.

At pressures below atmospheric pressure (that is, for
temperatures below the normal boiling point of the liquid) the

cohesive energy is the enthalpy of vaporisation, less the work done against atmospheric pressure. This quantity, when divided by the molar volume, provides the 'cohesive energy density', and the cohesion parameter is the square root of the cohesive energy density. It turns out that when two materials interact, we can assess the cohesive effects by multiplying the cohesion parameters together. (This gives a quantity with the dimensions of pressure: a kind of 'internal pressure' effectively holding the molecules together in the condensed phase, a quantity with some of the characteristics of a real, external pressure.)

The cohesion parameter approach to mixtures is that a material with a high cohesion parameter value requires more energy for dispersal than is gained by mixing it with a material of low cohesion parameter. In this case, the materials are immiscible. On the other hand, two materials with similar cohesion parameter values gain sufficient energy on dispersal to permit mixing. The method is attractive for practical use because it predicts the properties of a mixed system knowing only the properties of the components.

The simplest of the cohesion parameters is that proposed originally by Joel Hildebrand, the 'Hildebrand parameter'. We compile a list of liquids with gradually increasing Hildebrand parameter values to form a 'solvent spectrum' from which we deduce solubilities of materials such as polymers. In its most common form, it also includes subdivision into categories of hydrogen-bonding capability. By successive experimental choices, upper and lower pairs of two adjacent liquids in the list are identified, one of which dissolves the material and one which does not. This defines the solubility range in terms of cohesion parameters. Although specific ionic, polar and hydrogen bonding interactions can cause solvation effects, resulting in considerable 'scatter', the general trend is usually clear. We may obtain more precise information from multi-component cohesion parameters[10].

12.7 Gases in Liquids

Intuitively, one might think that all gases would dissolve in all liquids to about the same extent, but there are great variations.

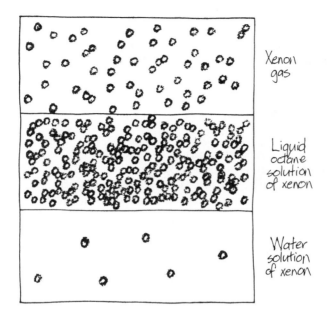

Figure 12.3 Schematic diagram of the solubility of xenon gas in octane and in water. Xenon is nearly forty times more soluble in octane than it is in water. In equilibrium with gaseous xenon, octane 'concentrates' the xenon, accommodating four times as many atoms as are present in the same volume of xenon gas phase at the same temperature and pressure. In contrast, water in equilibrium with xenon gas holds only one-tenth the number of xenon atoms as the xenon gas phase. After Pollack (1991).

Xenon, one of the noble gases, which you might expect to be fairly 'ideal' in its properties, is nearly forty times more soluble in the hydrocarbon octane than it is in water. In equilibrium with gaseous xenon, octane 'concentrates' the xenon, accommodating four times as many atoms as are present in the same volume of xenon gas phase at the same temperature and pressure. The aqueous phase holds only one-tenth the number of xenon atoms as the xenon gas phase (figure 12.3).

Even more striking, the solubility of one of the anaesthetic gases (halothane) in olive oil is a million times the solubility of common gases in the liquid metal, mercury[11]. (The reason for studying anaesthetics in olive oil is that olive oil is a 'model' for biological oils, and the anaesthetic effect depends markedly on the solubility of the anaesthetic in oily materials.)

12.8 Chemical Processes

Chemists seek models not only of materials but also of the ways they undergo interconversions. A model portraying such changes is a 'reaction mechanism', a detailed description of a reacting system as it progresses from reactants to products. In its most ambitious form, reaction kinetics includes consideration of reaction intermediates, the transition states and the interactions of all the reacting species with the solvent.

One aspect of the reaction mechanism model is the identification of all intermediates involved, the short-lived chemical species formed from the reactants which then undergo further reaction to provide the ultimate products. Another aspect is the assessment of the nature of the 'transition states', the high-energy states through which the reacting system passes in the rate-determining steps. The types of interaction of all species (reactants, intermediates, transition states) with the medium in which the reaction takes place, often a liquid solvent, can be crucial in controlling the reaction rate.

Physical chemists also employ the effect of a solvent on a chemical reaction rate as a 'probe' to investigate the structure of the solution. In this procedure, they use potential energy surfaces[12], with the reactants occupying one 'valley', separated from the products in another valley by 'mountains'. The lowest-energy route corresponds to a 'mountain pass', a route that sensible hikers would choose.

The foundation for the interpretation of chemical rate information is the 'law of mass action', first formulated in the middle of the nineteenth century. This says that the rate of any elementary process is proportional to the product of 'active masses' of the reactants involved[13]. At first it was believed that the **amount** of each of the reactants determined the rate of reaction. Then the model was improved to replace amount with the 'activity' or active amount, a concept which has its origin in the thermodynamic model. The thermodynamic model deals with conditions at equilibrium rather than the means by which equilibrium occurs or the time it takes. However, physical chemists invoke a modification of that model (the 'transition state' model), which assumes that there is a kind of equilibrium between the reactants and the species present in the transition state. It is then possible to apply the relationships of thermodynamics to rate or kinetic processes. Although this involves approximations, the results are useful. Associated with this model is the 'activation energy'. This is the minimum amount of energy accumulating within a chemical bond or group of bonds before a successful reaction can result. The potential energy surface model assesses this energy from the height it is necessary to climb from the reactant 'valley' to the highest point on the 'mountain pass'[14].

Another concept used in chemical reaction kinetics is the 'relaxation time'[15]. Each chemical process (which is a 'relaxation' towards equilibrium) has a characteristic time scale described by the relaxation time. The time scale of any method used to investigate a reaction has to be of the same order of magnitude as the relaxation time. If we watch a chemical system over a period of hours without any special instrumentation, we cannot 'see' any process which is complete in less than a millisecond. Nor are we able to see a reaction that will not be complete for years. The possible time scales in chemical rates of reaction are as diverse as in other areas of science. They range from times of the order of a femtosecond (a millionth of a nanosecond) to periods longer than Earth's geological age. Improved chemical instrumentation enables us to follow very rapid chemical processes, but not the very slow ones. However, computer simulation can now provide an indication of the mechanisms of reactions not directly observable[16].

There is an exponential relationship[17] between the activation energy for a process and its relaxation time, so small variations in energy between different reactants can result in dramatically different relaxation times. Another factor is that the chemical processes themselves do not necessarily control the overall rates of the faster chemical reactions. Instead, the rate is limited by how fast the reactants can diffuse together, or how fast the products can diffuse away after the reaction.

12.9 Catalysts and Enzymes

An important aspect of chemical reaction kinetics is the development of materials to serve as 'catalysts' to speed up the slow reactions, to facilitate reactions that would otherwise be inconveniently slow. Catalysts may be homogeneous (in solution with dissolved reactants) or heterogeneous (providing a surface on which the reactants adsorb). There are also intermediate situations, with catalyst groups which are 'in solution' but tied to insoluble polymers, or catalysts associated with surface-active agents (surfactants) such as detergents in micelles or microemulsions. These situations provide large numbers of local environments suitable for reaction within a homogeneous phase. An extension of this idea is the microemulsion-based gel[18].

Enzymes in biological systems are the ultimate catalysts, achieving rates of reaction and providing highly specific products beyond the scope of 'synthetic' catalysts. The search for better industrial catalysts draws heavily on the study of enzymes, concentrating on the synthesis of organic molecules having functional groups with defined orientations.

The great efficiency of enzymes as catalysts is attributed to energetically favourable interactions between reactants in an 'encounter complex'. The encounter complex model proposes an assembly of reactant and enzyme molecules in intimate physical interaction, but without any specific chemical bond formation. It locks all the components into the spatial orientation necessary for the reaction to continue through the high-energy transition state and onto the products. This physically bound encounter complex in enzyme systems takes the place of the metastable chemically bound intermediates that we often observe in other forms of catalysis. In the encounter complex there are several possible stabilising contributions

- specific favourable physical interactions such as hydrogen bonding between particular groups,

- an environment that does not 'solvate' or 'deactivate' reactive groups on the reagents as an aqueous solvent would, for example. (This effect can be imitated by carrying out reactions in a 'polar aprotic' or 'proton-free' solvent like acetonitrile.)

- Increased favourable dispersion or London interactions. (Dispersion interactions are greater if the matter is denser, because there are more electrons per unit volume.)

All three effects can occur if a large enzyme molecule wraps itself closely around the reaction site so that the atoms are packed intimately together. Most atoms are only short covalent **bond** distances apart, rather than the longer **intermolecular** distances they would have if small solvent water molecules surrounded the reaction site. As a result, specific favourable interactions can occur, specific unfavourable interactions are prevented, and there are more dispersion interactions. Recent studies of catalysis by both enzymes and nonenzymatic proteins like albumin have emphasised the significance of a protein 'pocket' as a suitable model for the environment provided for such chemical transformations[19].

The thermodynamic model tells us that the encounter complex is stabilised in terms of **energy** relative to the separate reactants. This occurs despite a considerable **entropy** disadvantage because the encounter complex is a very ordered structure. All the reactants temporarily attach themselves to the same structure, and arrange themselves in the correct positions and orientations for the reaction. The final reaction can then take place rapidly. The encounter complex is very short-lived, with reaction rapidly continuing through the transition state to yield products and release the enzyme for another cycle of reaction. The turnover rate is high: typically a thousand molecules of reaction per enzyme molecule in every second.

Although our synthetic catalysts are inferior to enzymes, they do have one advantage over biological processes. Unlike biological processes, which are limited to aqueous systems, we can choose the solvents. There is a wide range of pure liquids and their mixtures as potential solvents. We balance two effects in choosing a solvent as a medium in which to carry out a reaction. The interactions the solvent has with the reactants must be

- strong enough for both reactants to dissolve, but

- not so strong that solvation of the reactants inhibits the desired reaction.

Alternatively, we can use a two-phase system ('phase transfer catalysis') or use microemulsions to provide local environments favourable to reaction and with large surface areas exposed to the reagents[20].

We can arrange to control some chemical reactions by application of an electrical potential in the process of 'electrolysis', where adjustment of the applied electrical potential provides fine-tuning of the reaction conditions. If the products and/or reactants are coloured there is now the technique of 'electrochromism', with the opportunity for a variety of new applications[21].

Endnotes

[1]See also **13.8 Solid Solutions**, **5.8 Phase Changes** and **5.9 Dynamic Models**.

[2]Amount of substance has units of moles; it is not the same as mass but is defined in **9.1 Amount**. See also **10.5 Ideal Gases**.

[3]Further reading Lovelock (1988, p 18).

[4]See **3.1 States**.

[5]See **10.5 Ideal Gases**.

[6]Further reading Tanaka (1981). See also **15.5 Gels**.

[7]See **12.4 Phase Diagrams** and **2.4 Condition Critical**. Critical points also exist in mixtures of liquids (at the temperature where miscibility just occurs), in ferromagnetism (**13.9 Magnetic Matter**) where there is a critical temperature called the Curie point, in superconductors (**13.7 Superconductors**), in superfluids (**11.3 Superfluid Helium**) and in polymeric solutions (**15.3 Polymer Solvents**), gels (**15.5 Gels**) and metal alloys. Further reading Wilson (1979), Scher *et al* (1991), Gleick (1988), Cardy (1993).

[8]Further reading Walker and Vause (1987), who used 'Reappearing Phases' as their title.

[9]See also **3.7 Chaos**. The mixing of viscoelastic fluids that tend to return to their original shapes after deformation (see **11.4 Viscous Flow** and **15.4 Flowing Polymers**) involves additional complicating factors. Further reading Ottino *et al* (1992), Ottino (1989).

[10]'Cohesion parameters' were previously called 'solubility parameters'. I introduced the new term to emphasise the wide range of applications of this model. Further reading Barton (1975, 1983a, b, 1987, 1990, 1991).

[11]Further reading Pollack (1991).

[12]Described in **5.1 Energy Well Model**.

[13]See **9.1 Amount**.

[14]Described in **5.1 Energy Well Model**.

[15]See **2.2 Perspective** and **3.2 Dynamics**.

[16]Further reading Gillan (1993), Suzuki *et al* (1992), *Physics Today* (1993). See also **10.4 Computer Simulation**.

[17]Described in **2.2 Perspective**.

[18]See also **14.4 Adsorption** and **14.7 Surfactants**. Further reading Rees and Robinson (1991).

[19]See **10.1 Molecular Interactions**. Further reading Benkovic (1996) and Hollfelder *et al* (1996).

[20]**14.7 Surfactants**. Further reading Rees and Robinson (1991).

[21]Further reading on electrochromism Silver (1989).

CHAPTER 13

ORDERED MATTER

Crystalline solids have characteristic electrical and magnetic properties, as well as strength and stability. It is usual to classify the materials as insulators, conductors, semiconductors, superconductors or ferromagnetics. We can discuss all these attributes in terms of simplified models of solids and extend the concepts to real materials with similar properties. Pictorial analogies can illustrate the various types of solid. For example, we can model the units in a light-hearted way on our own social situations[1]†:

● molecular crystals (each molecule modelled as a couple), or

● oppositely charged ions in an ionic crystal (unpaired men and women), or

● metal ions and electrons in metals (an array of people within a swarm of bees).

Rather than using existing bulk materials of uncertain composition and configuration, some physicists prefer to grow their own.

13.1 Layer by Layer

Physicists and chemists apply the processes of 'molecular beam epitaxy' and 'metal organic vapour phase epitaxy' to build up a

† Endnotes for this chapter can be found on page 293.

material in successive molecular layers. (Semiconductors for microlasers and magnetic multilayers are important applications at present.) They place the substrate and containers of each of the required chemical elements in a vacuum chamber. As they heat each container, that material evaporates and flows at a controlled rate through an opening. For example, they would make a layer of a compound such as gallium arsenide by allowing the sources of gallium atoms and arsenic atoms access to the substrate for a certain time. They would then use a scanning tunnelling microscope to study the resulting surface structure on an atomic level. Other scientists might use electrodeposition instead, avoiding the interdiffusion between layers that may occur at high temperatures[2].

Thus we have freed ourselves (in principle) from a reliance on nature for many pure and doped crystals. However, most crystalline materials in current use are either obtained directly from natural sources or formed by relatively unsophisticated technologies. Synthesising crystals is like designing a building, where we use specified shapes and sizes of bricks or planks or girders and decide how we will fasten them together[3].

13.2 Crystals

Crystalline solids are characterised by an orderly, cohesive, close-packed, periodic arrangement (a 'crystal lattice') of molecular particles (atoms, ions or molecules). They have a very ordered internal structure that produces reproducible, recognisable external crystal faces. The thermodynamic model favours a highly ordered configuration despite the entropy penalty because of the great stability of the resulting chemical bonds. Although the molecular particles have vibrational kinetic energy, this is usually inadequate to enable them to exchange positions except under certain conditions near the crystal surfaces. As I have already shown[4], we can study the internal structure by using the diffraction of electromagnetic radiation with a wavelength comparable to the particle spacing.

Crystalline solids are the most ordered form of matter, but it is still helpful for us to use simplified models of them. To start with, crystallographers (scientists specialising in crystal structures) assume ideal crystals, without irregularities, although real macroscopic crystals have high levels of imperfections. They then concentrate on the smallest element of crystal and repeat it indefinitely to make up the whole—like the repeat distance of the pattern on a wallpaper, but in three dimensions. This is the 'unit cell'.

The idea of symmetry is particularly important in crystal structures. A crystal is an assembly of identical molecular particles. Because this molecular particle on its own usually has a low symmetry, they call it the 'asymmetric unit'. They then look for several asymmetric units that together make a symmetrical configuration, described mathematically by a 'space group' of which there are only a certain number possible. They define the unit cell in their model of the crystal as the 'box' that encloses the set of asymmetric units (molecules) described by the space group[5].

The simplest type of unit cell is a cube (with all sides of equal length), creating a cubic lattice, with the centre of an atom at each corner of the unit cell.

We can imagine the unit cells repeated in three dimensions to build up the crystal lattice. The three common cubic structures are

- 'simple cubic' (one-eighth of a particle at each of the eight corners of the unit cell),

- 'body-centred cubic' (one-eighth of a particle at each corner plus one centred), and

- 'face-centred cubic' (one-eighth at each of the eight corners plus one-half on each of the six faces of the unit cell).

Of these, the face-centred cubic packing of identical spheres uses space most efficiently (74%). Another packing method providing the same density but with hexagonal rather than cubic symmetry is

● hexagonal close packing.

In both cases each site has twelve nearest neighbours, but those metals with face-centred cubic packing are malleable and ductile because the planes of atoms can move past each other more readily[6]. In the packing of nonspherical molecules, there is a minimum of free space (typically 65 to 77% of the space being occupied) because the molecules arrange themselves to maximise van der Waals interactions[7].

You can think of unit cells as bricks and the crystals as buildings. Real crystals built from unit cells do not look exactly the same as each other any more than buildings built from the same bricks look the same. The relative growth rates in different directions as well as the nature of the unit cell determine the ultimate crystal shapes.

There are only certain ways that identical unit cells can fit together without leaving 'gaps', which is the reason a crystal cannot have fivefold symmetry. Dodecahedral unit cells (structures with twelve faces, each a pentagon) cannot completely fill space when packed together. Note, however, that there is an alloy of aluminium, copper and iron demonstrating a state of matter made up of 'quasicrystals'. Here there are dodecahedral grains that do have fivefold symmetry and therefore cannot be perfect crystals. We can consider quasicrystals from several points of view. Roger Penrose of Oxford University used a combination of two or more different unit cells. We can also model quasicrystals as a form of glass with ordered clusters and defects, and most recently a 'random tiling' model ('Penrose tiling') has been used. We also have to be more flexible in our attitudes to dimensions, because we find it is possible to model these three-dimensional structures as projections from a six-dimensional 'hyperlattice'[8].

Another point to remember is that real crystals, unlike ideal or model crystals, are imperfect. When a crystal breaks, the fracture tends to follow defect lines or planes[9]. You should note also that the equilibrium surface for any crystalline plane on the thermodynamic model is not perfect. This is too highly ordered, and to minimise the free energy in the thermodynamic model the system achieves a balance between energy and entropy effects. Therefore there will be incomplete planes, edges, dislocations and vacancies. Experimentally, surface scientists can observe these surface dislocations with relatively simple equipment. When they etch a crystal in a solvent, they see under a microscope the etch pits where preferential dissolution from 'high-energy' sites has occurred. The presence of defects and dislocations greatly modifies the mechanical performance of pure materials[10].

For convenience, crystals are often classified formally as follows, although these classes represent ideal or model behaviours rather than sharp distinctions between various kinds of real crystal:

- covalent crystals, with lattice sites occupied by atoms, as in diamond,

- noble gas crystals, with van der Waals interactions between sites,

- ionic crystals, with coulombic interactions between sites,

- molecular crystals, with lattice sites occupied by molecules,

- metals, with sites occupied by metallic ions.

Chemists and physicists use x-ray crystallography extensively for investigating the molecular structure in three dimensions of large organic molecules and macromolecules. It is therefore of great practical importance to be able to 'grow' molecular crystals of these compounds. This is particularly valuable for proteins and nucleic acids, providing unique structural information. The limitation of this method is that most of the larger biological molecules do not crystallise readily. The synthesis of molecular crystals is receiving attention with the aim of obtaining materials with particular electronic, optical or magnetic properties[11].

An interesting footnote on crystalline solids is the formation of 'petrified' objects or 'fake rocks' in natural springs caused by bicarbonate decomposing to carbon dioxide and carbonate.

Carbonate precipitates out with calcium ions as travertine, a form of calcium carbonate, on any objects in the water[12].

13.3 Nucleation

The molecular particles (molecules, atoms or ions) on the surfaces of crystals cannot make as many stabilising interactions with neighbours as those in the interiors of crystals. Therefore very small crystals, with high surface-to-volume ratios, are much less stable than macroscopic crystals. As a result, larger crystals grow at the expense of smaller ones. The problem is to get the first small crystal. There is a 'bottleneck' in the crystallisation process associated with the formation of a 'nucleus' (the initial array of particles in the correct arrangement for subsequent growth), the process being known as 'nucleation'. (The word 'nucleus' is being used here with an entirely different meaning to that in an atomic 'nucleus'.) The result is that liquids and solutions can remain free of crystalline solids under conditions where in the thermodynamic model one would expect crystals. They exist in a metastable state[13].

If water is pure enough, stable ice nuclei will not form even at minus 20 degrees Celsius, well below the normal freezing point. Droplets of water that condense inside a cloud may not freeze even at temperatures as low as minus 40 degrees Celsius. Some other materials (such as silver iodide and long-chain alcohols) have crystals similar in shape to those of water and can nucleate the freezing process. Scientific rainmakers use these to 'seed' or nucleate clouds, because the 'latent' heat[14] released by crystallisation warms the cloud, which then rises and more water vapour condenses. Nucleation of supersaturated water vapour in the atmosphere, for example by sulphuric acid and other sulphur compounds, has an important climatic effect. Some crystals (such as amino acids) that have little structural similarity to water crystals also nucleate water, and this we attribute to high electric field gradients in small cracks within the crystals. The presence or absence of chirality in the crystals is also significant in some cases. Crystals and droplets are not the only nucleation initiators.

Pseudomonad bacteria can induce freezing in water droplets, and the nucleation of supersaturated calcium carbonate by organisms is an important geophysiological process[15].

13.4 Electrical Conduction

Isolated metal atoms are reactive, and have a tendency to form metal–metal bonds if they are not in a chemical environment where they are able to react with nonmetals to form ionic compounds. As with all condensed matter, cohesive energies provide a useful measure of the strength of metal–metal bonds, although small numbers of metal atoms grouped together (microclusters) have properties different from those of bulk samples.

The simple electrical conductivity classification model of substances described in Chapter 9 is probably familiar, particularly the electrical conduction properties of metals. We often describe them in terms of a model in which the electrons have relative freedom of movement through a lattice of positive ions. There is also increasing interest in those organic materials that have electrons free to carry current.

In the molecular orbital model, bonding electrons are free to distribute themselves throughout the whole of a molecule rather than to be localised in a chemical bond. This concept leads on to the more extensive electron delocalisation in the 'band model' of solids. In the formation of molecular orbitals, we assume that the atomic orbitals of individual atoms combine to provide bonding orbitals with energies lower than the original. There are also antibonding orbitals with higher energies. As more and more atoms participate in this process, there is a continued splitting of orbital energy levels, until there is an effectively infinite number of very closely spaced allowed energy levels within a band. Because of the energy separation of atomic orbitals, more than one discrete energy band forms in the resulting solid, with an energy gap between them. In metallic solids the lowest band is half-filled, and electrons in orbitals within the band readily acquire thermal energy to occupy empty orbitals and so are very mobile. We can interpret the effect of a temperature rise as increasing the extent of 'collisions' between the moving electrons and the atoms in the lattice, making the electrical conduction less efficient.

In insulators and semiconductors, electrons fully occupy the lower or valence band, and conduction can occur only if we promote electrons across the band gap to the conduction band. In a semiconductor, electrons are thermally excited as the temperature increases, but in an insulator the band gap is too great.

In this model, if physicists 'dope' a solid lattice with a chemical element that can trap electrons (for example, the doping of arsenic with germanium atoms) there are 'positive holes' left. These holes permit the remaining electrons to move, providing 'p (for positive)-type semiconductivity'. Alternatively, doping with atoms with excess electrons provides electrons in an otherwise unoccupied band that can cause 'n-type semiconductivity'.

13.5 Organic Conductors

If electrons are delocalised throughout the length of a macromolecule there is the opportunity for polymers to be electrically conducting as long as the macromolecules are appropriately aligned. For example, polyaniline appears to be heterogeneous, with metallic domains of high electrical conductivity. The metallic properties of polyacetylene are enhanced by doping to give a material with twice the electrical conductivity of copper, weight-for-weight[16].

One of the fascinating properties of the electrically insulating fullerenes is their ability to form compounds with metals. The process of forming these 'intercalation' compounds we sometimes also describe as 'doping' with metal atoms. The intercalation compound K_3C_{60} has a lattice of C_{60} molecules, with potassium atoms fitting tidily into the free spaces, an arrangement common in ionic crystals. This compound has good electrical conductivity. The potassium atoms donate electrons to the fullerene molecules, forming an ionic material with a large cohesive energy, and the electronic properties are due to the electron-rich fullerene ions, not to the potassium ions. This compound becomes superconducting below 18 kelvin. Other compounds such as K_6C_{60} are not superconducting[17].

It is also possible to internally dope the fullerene cage. The 'space' that exists in an 'empty' buckminsterfullerene molecule is large enough to accommodate almost any other atom (the lanthanum compound was the first synthesised). Alternatively, we can incorporate an atom such as boron or nitrogen in place of carbon in the molecular structure itself[18].

13.6 Ceramics

Ceramics are solid materials in which metallic elements combine with nonmetallic elements, usually oxygen, in particular structural arrangements that make them 'refractory' (high-melting) and hard. In their 'ideal' state they are brittle, but we can toughen them by incorporating pores or voids that intercept the propagation of cracks. An example is the plasma-sprayed coating formed when one introduces ceramic powders into a plasma 'flame'[19].

The behaviours of electrons within ceramics determine whether these materials are insulators or conductors. In a compound like barium oxide we consider the electrons to be localised in the regions around the barium and oxygen nuclei. Oxygen atoms capture the higher-energy outer electrons of the metal to form an ionic bond. Consequently, this type of ceramic is an electrical insulator. In contrast, many copper oxide ceramics are good conductors if other chemical elements are present. This is because copper and oxygen share electrons in a covalent bond and some of the electrons are not localised near particular atoms. In addition, some particular combinations are superconductors[20].

The perovskite family of ceramics is particularly interesting because in their electrical properties they range all the way from insulators to semiconductors, to metal-like conductors and even to superconductors. Perovskites have the general formula ABX_3, where A and B atoms are positive metal ions (cations) and X atoms are negative nonmetallic ions (anions). The compounds that conform to the 'ideal' perovskite structure are insulators, with all atomic sites filled and strong attractive interactions between the cations and anions. They are isotropic, with uniformity of properties along each axis of the unit cell. However, perovskites may be distorted if the relative sizes of A and B cations cause some of the atoms to move out of the 'ideal' positions. Off-centre cations give these 'tilted' perovskite crystals electrical polarity, one end negative and the other positive. If we can change the direction of off-centring by applying an electric field, the material is a 'ferroelectric', with potential applications in electronic devices. Barium titanate ($BaTiO_3$) is such an electroceramic, storing electrical energy and then slowly releasing it. (This has application in protecting circuitry, for example to protect computers during a lightning surge.) Barium titanate also has 'piezoelectric' properties (electric charge separation associated with mechanical stress). When the titanium atoms shift in an electric field the whole crystal changes shape, so this material can act as a transducer converting mechanical energy to electrical energy or vice versa, as in speakers or microphones. Perovskites with mixed cations (solid solutions of various compositions) can act as semiconductors. Also, the perovskite including yttrium, barium and copper ions is a new high-temperature superconductor[21].

Many of these materials are metastable[22]. Another material resulting from the crystallisation of metastable glasses is a 'glass ceramic' that has most of the properties of glass but is opaque. All glasses eventually crystallise (devitrify), although the time required may be far too long on a human scale. Some glasses do crystallise in a conveniently short time, and result in a crystal that is very strong because the dislocations are unable to move.

13.7 Superconductors

In a metallic conductor we model the electrons as occupying an 'energy band'. In a superconductor an attraction between electrons results in their condensation into a low-energy state separated by a gap from the 'normal' energy band. The phenomenon was observed by Heike Kamerlingh Onnes in 1911 and named 'superconductivity' by 1913. The first material observed to have a phase transition to a vanishingly small electrical resistivity was mercury at temperatures below 4 kelvin (minus 269 degrees Celsius, 4 degrees above absolute zero). A famous demonstration took place in 1932. In Leiden, current flow was initiated in a lead ring (cooled by liquid helium) which was then flown to London with the current still flowing for a demonstration at a Royal Institution lecture[23].

There followed the investigation of related phenomena, including the expulsion of small external magnetic fields (the 'Meissner effect', so named because of the work of Walther Meissner in the 1930s). When cooled to below the transition temperature in the presence of a magnetic field, the magnetic flux is expelled, leaving the superconductor as a perfect 'diamagnet' (a material repelled by both poles of a magnet). This is what enables the popular demonstration of a superconducting material being levitated above a magnet. It also allows superconductors to act as shields against magnetic fields, just as ordinary conductors shield against electric fields.

Researchers then sought materials showing superconducting properties at more accessible temperatures, in particular above 77 kelvin (minus 196 degrees Celsius). Materials that become superconductors above this temperature can operate while being cooled by the readily available and inexpensive liquid nitrogen rather than by liquid helium. Not only are the running costs considerably lower, but the complex and expensive helium vapour recovery systems are unnecessary.

The report of superconductors at 'high temperature' (temperatures high compared to the boiling point of liquid nitrogen) among materials based on copper oxide ('cuprates') in 1986 caused a dramatic expansion in these efforts. Researchers have made significant progress, although the promises of practical applications have become more conservative. High currents through superconducting coils can provide light, powerful

magnets without iron cores, but a practical ambient temperature superconducting power transmission line is still beyond reach[24].

The accepted general model based on quantum mechanics centres on the formation of bosons, particles not subject to exclusion rules. When cooled to the transition temperature for the particular superconductor, electrons pair up as 'Cooper pairs' (named after Leon Cooper who worked at the University of Illinois in the mid-1950s). In an ordinary metal we 'explain' the electrical conductivity as mobility of electrons, which behave as fermions. However, in the superconducting phase the electrons change their characteristics by pairing up and (in terms of the quantum mechanical wave function model) becoming bosons. As the material cools into its superconducting state (without an applied magnetic field) the phase of the wave function becomes macroscopically correlated instead of varying throughout the material. All electron pairs are in the same quantum state, so there is no mechanism for energy dissipation. It is as if the electrons undergo a phase transition to a structure with long-range order[25]. We can also use the model of composite bosons to 'explain' other electrical and magnetic effects in highly conducting materials[26].

Questions in the course of being answered for each potential high-temperature semiconductor include experimental ones:

- Is the particular material a 'true' superconductor?

- Is there a phase transition to a zero-resistance state, or does the resistivity just become too small to measure?

- To what extent does this depend on the strength of the external magnetic field?

There are also theoretical questions concerning the 'mechanisms' for the effects described by the models. As well as the higher-than-usual temperatures for transition to superconductivity, the 'normal' metallic states of these materials are also interesting.

Although superconductors expel small magnetic fields, intense fields penetrate them. In 'type I' superconductors small external magnetic fields are expelled from the superconductor, but higher-strength fields penetrate and destroy the superconductivity. However, in 'type II' superconductors although small magnetic fields are expelled, those in excess of a critical value penetrate nonuniformly, forming a 'mixed state'. In the present model, the magnetic field is admitted in the form of quantised vortices[27].

In a superconducting material the model assumes external magnetic field 'flux lines', each carrying one quantum unit of magnetic flux (associated with one Cooper pair of electrons). Electron interference micrography has revealed flux lines 'dancing across the surface of a superconductor', with interference fringes occurring one flux quantum apart. The flux lines are associated with two characteristic lengths. One is the 'penetration depth' defining the diameter of each flux line. The other is the 'coherence length' describing the radius of the vortex core where the amplitude of the Cooper pair wave function is zero. The coherence length indicates the spatial extent of the superconducting pair, the shortest superconductor length over which the electrical resistivity can change, and of the order of tens or hundreds of nanometres. The penetration depth is inversely proportional to the density of superconducting pairs and indicates the distance which a magnetic field will penetrate. This is the shortest superconductor length over which the magnetic properties can change.

The value of the ratio of the penetration length to the coherence length differentiates type I from type II materials. The former have coherence length greater than penetration depth, so the superconducting state disappears at relatively low magnetic field strengths. In type II superconductors the penetration depth exceeds the coherence length, so the superconductivity remains even in the presence of a magnetic field. In superconducting materials, physicists model stable electric supercurrents as flowing around the vortex, screening the magnetic field and confining it to within the penetration length. Vortices can exist either as a regular lattice with long-range order in a vortex lattice, or with less extended order in a vortex glass. Experimentally, one can obtain images of 'vortex patterns' with and without long-range order from fine magnetic particles on the surfaces of superconducting crystals. This mirrors the way we can see magnetic field lines from iron filings on a magnet. We can also obtain images from scanning tunnelling microscopy and electron holography.

In high-temperature type II materials in the presence of a magnetic field there is also a vortex fluid state at temperatures above the 'freezing point' of the vortices. Here the vortices are mobile and with only short-range correlations. Physicists construct phase diagrams, with axes of applied magnetic field and temperature, to depict transitions between the various states. In type I materials at low fields the magnetic field and the vortices are expelled, and there are no supercurrents. We say that the material is in the 'Meissner phase'. The properties of 'low-temperature' and 'high-temperature' type II superconductors differ markedly. The high-temperature type II superconducting materials like cuprates with penetration lengths much larger than coherence lengths are extreme type II superconductors. In these materials, as the temperature decreases in the presence of a magnetic field there is at first no well defined transition but a decrease of resistivity attributed to gradual Cooper pairing of electrons, until a temperature is reached where there is an abrupt transition to a true superconducting phase. This is in contrast to the behaviour of a low-temperature type II superconductor with just a sharp transition[28]. There are other interesting phenomena in superconductors only apparent at ultralow temperatures: millikelvin, microkelvin and even nanokelvin temperatures[29].

There are close parallels in the models used to describe magnetic properties such as ferromagnetism, below, and those used for superconductors. The models currently in use to describe superconductivity still fall far short of fitting all observations, and much of the experimental activity is still at the empirical level[30].

One of the ceramics, yttrium(1)–barium(2)–copper(3) oxide (sometimes known as 1-2-3 or YBCO), superconducts at 90 kelvin (minus 183 degrees Celsius). Two other copper oxides incorporating bismuth on the one hand and thallium on the other show these properties at even higher temperatures. The currently preferred model of the cuprate superconductors that arises naturally from the structural characteristics is the 'charge-transfer' model. This uses stacks of closely spaced conduction layers of CuO_2 in which the superconductivity occurs, separated by charge reservoir layers that provide the carriers or the mechanism necessary for superconductivity. The nature of the conductivity depends on the amount of charge transferred between the conduction layer and the charge reservoir layer. This is controlled by the available oxidation states of the metal atoms in the reservoir layer. The large range of variations possible in the components and spacings provides the opportunity for a range of properties but also requires extensive study. Existing models have insufficient detail to accommodate all experimental results, but they provide a broad correlation of behaviour, and intuition and luck have proved equally important in isolating superior materials[31].

The cuprates are clearly anisotropic (properties along directions within the layers or planes differing from those perpendicular to the planes) so high-quality crystals are necessary for precise measurements. The electrical resistivity varies by factors of up to a hundred thousand according to direction, and its nature is different. The in-plane resistivity is similar to that of a metal, while in the perpendicular direction it is more like that of a semiconductor (increasing when the temperature rises).

From the point of view of superconductivity, in-plane behaviour is of most interest. It is possible to sketch general or 'generic' phase diagrams for layered cuprates by mapping temperature against electron concentration in the CuO layers (controlled by adjusting doping levels). At low doping levels the compound is an insulator with 'antiferromagnetic' long-range magnetic order (see **13.9 Magnetic Matter**, below). As the electron concentration increases, the material becomes an insulator, with only short-range order. At higher electron concentrations it is a metal with a transition to a superconducting ground state, and at still higher levels it is a normal metal with electrical resistivity that decreases smoothly with decreasing temperature[32]. The metastability of oxide superconductors influences their synthesis, including the extent of defect concentration, but once prepared they do not tend to decompose into more thermodynamically stable phases[33].

A newly developed group of superconductors comprises the doped fullerenes mentioned above. In these molecular solids, the molecular units do not share electrons, unlike the high-temperature ceramic superconductors which have a continuous ionic lattice. The 'metallic' K_3C_{60} may be suitable for practical superconducting wires, although we must protect it from the air because the potassium atoms are still reactive to oxygen[34].

13.8 Solid Solutions

Compositions of solid ionic crystals frequently depart from the integral stoichiometric ratios[35]. Therefore chemists prefer not to describe these materials as 'compounds', this word suggesting strict stoichiometry as a result of specific bonds between atoms.

As long as there is an overall charge balance, and the organisation of atoms of different sizes satisfies space requirements, 'whole numbers' are not important in these arrays. Examples include the oxide materials that are variations on the 'ideal' stoichiometric compounds (such as $YBa_2Cu_3O_7$) with some defects and some metal atoms replaced by others. In some situations it is easier to treat such a non-stoichiometric material as a stoichiometric compound with minor defects. However, if the deviations from stoichiometry become significant the solid solution model is preferable.

13.9 Magnetic Matter

Think of electrons in a solid as behaving as tiny magnets because of their quantum mechanical spinning. Then in a ferromagnet,

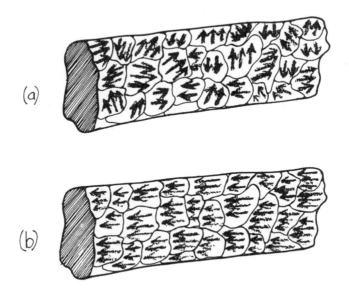

Figure 13.1 (a) Ferromagnetic domains in unmagnetised iron are randomly arranged. (b) During the magnetisation process, the domains line up, producing opposite poles at each end of the bar. (The diagrams are not to scale.

visualise the interactions between electron spins causing them to align cooperatively in the same direction throughout the material in a long-range ordering process[36]. An ordinary 'unmagnetised' sample of the ordered, ferromagnetic iron does not show magnetic properties. This ordered state occurs as randomly oriented domains until the sample is 'magnetised'. However, **within** each domain the there is ordered orientation (figure 13.1). 'Ferrofluids' contain magnetic particles that are small enough not to contain more than one domain, so when we apply a magnetic field, they align with it.

The thermodynamic model tells us that ferromagnetism occurs in iron, cobalt, nickel and other materials because there are energetic advantages for these materials in aligning atomic spins in this way. We must provide energy to the system to disrupt the order. Thermal energy has this effect, and in iron above 771 degrees Celsius the directions of the magnetic moments become random and the iron is paramagnetic. This is the 'Curie point'[37].

In 'antiferromagnetic' materials like chromium the electron spins show order, but in their low-energy states there is a periodic spatial structure in their orientation rather than a uniform orientation. For example, neighbouring atoms align in opposite directions, so there is no net effect in the presence of a magnetic field (figure 13.2). There is a critical temperature at which antiferromagnetic behaviour changes to paramagnetic behaviour.

'Magnetic multilayers' are metallic magnetic superlattices formed by the deposition of successive films of two or more metals, at least one of which is magnetic. These materials provide a range of interesting physical properties with technological potential[38].

Related materials are 'spin glasses', the name being based on two distinct models. The first part, 'spin', means the quantum mechanical spin. The other part, 'glass', refers to the disorder in the configurations of those spins, analogous to the lack of long-range order in an ordinary glass. These materials, for example a copper lattice containing some iron atoms, display both ferromagnetic and antiferromagnetic properties. The nature of the alignment depends on the separation of the pairs of atoms, with

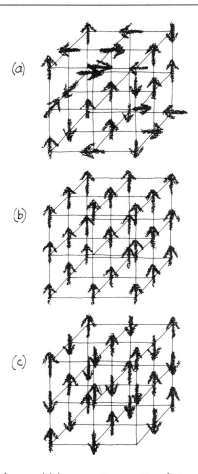

Figure 13.2 If we model the magnetic properties of atoms as arrows, in paramagnetic materials (a) the arrows are randomly oriented, in ferromagnetic materials they are aligned (b), and in antiferromagnetic materials (c) adjacent atoms are aligned parallel but in opposite directions.

parallel and antiparallel alignments being alternately preferred. The result is that atoms cannot satisfy all the requirements for minimum energy simultaneously, a situation described by the colourful term 'frustrated systems', and there are many possible 'low-energy' states. There are similarities between the spin glass phase and the vortex glass phase in superconductors[39].

Endnotes

[1]Fortman (1993).

[2]Further reading Jewell *et al* (1991, p 59), Main (1993), Falicov (1992), Chang and Esaki (1992), Jones and McConville (1995), Cahn (1996). See also **7.7 Quantum Dots**.

[3]Further reading Fagan and Ward (1992).

[4]**6.2 Probing Matter** and **6.3 Transformation**.

[5]Further reading McPherson (1989), Heilbronner and Dunitz (1993, Chapter 9). See also **2.8 Symmetry**.

[6]Further reading Selinger (1989). Cubic packing is illustrated in figure 13.2 and hexagonal packing in figure 6.4.

[7]See **10.1 Molecular Interactions**. Further reading Fagan and Ward (1992).

[8]See **2.6 Limits and Dimensions**. Fivefold symmetry also appears in fractals, **3.8 Fractals**, such as aggregates (**14.13 Aggregates**) like soot particles generated by random events. Further reading Stephens and Goldman (1991), Elliott (1991), Heilbronner and Dunitz (1993, p 115).

[9]Further reading Finnis (1993).

[10]See **5.4 Energy and Entropy** and **6.5 Failure**.

[11]Further reading McPherson (1989), Fagan and Ward (1992).

[12]Further reading Pentecost (1991).

[13]See **5.9 Dynamic Models** and **14.1 Windows**. There is a story about an early organic chemist whose success in crystallising reaction products was attributed to the great variety of tiny crystals which he had involuntarily accumulated in his beard and which acted as nuclei as they fell into the solutions while he worked.

[14]See **5.8 Phase Changes**.

[15]Further reading McBride (1992).

[16]See also **9.13 Aromatic Model** and **15.2 Polymers**. Further reading Jeon *et al* (1992), Flam (1991), Kaner and MacDiarmid (1988), Corcoran (1990, p 79).

[17]Further reading *New Scientist* (1991), Zhou *et al* (1991), Curl and Smalley (1991, p 40), Chen *et al* (1991), Uemura *et al* (1991), Stephens *et al* (1991), Hammond (1992), Hecht (1991b), Smalley (1991), Hebard (1992). See **13.7 Superconductors**.

[18]Further reading Hammond (1992).

[19]See also **8.11 Plasmas**. Further reading Herman (1988).

[20]See also **9.3 Bond Models** and **9.7 Electronegativity**, and **13.7 Superconductors**. Further reading Cava (1990).

[21]Further reading Hazen (1988), Sleight (1991).

[22]A property discussed in **5.9 Dynamic Models**.

[23]Further reading Donnelly (1995).

[24]Further reading Rowell (1991), Wolsky *et al* (1989), Cava (1990), Bishop (1993).

[25]See also **4.1 Ultraviolet Catastrophe, 4.5 Playing Cards and Four-Strokes, 11.3 Superfluid Helium, 4.2 Wave States, 3.7 Chaos** and **5.8 Phase Changes**.

[26]Further reading Kivelson *et al* (1996).

[27]See also **11.3 Superfluid Helium**.

[28]See also **8.10 Surface Images** and **12.4 Phase Diagrams**. Further reading Ball and Garwin (1992, p 764), Raveau (1992, p 56), Hebard (1992), Bishop (1993), Bishop *et al* (1993).

[29]Pobell (1993).

[30]Further reading Huse *et al* (1992), Wolsky *et al* (1989), Bardeen (1990). See **13.9 Magnetic Matter**.

[31]Further reading Batlogg (1991), Jorgensen (1991), Cava (1990), Girvin (1992), Raveau (1992), Tilley and Tilley (1990).

[32]See diagram, Batlogg (1991, figure 2). See also **13.9 Magnetic Matter**.

[33]Further reading Sleight (1991).

[34]Further reading Curl and Smalley (1991, p 40), Hebard (1992).

[35]See **9.1 Amount**.

[36]See **3.7 Chaos** and **4.4 Heisenberg May Have Been Here!**.

[37]Curie temperatures are critical points (**12.4 Phase Diagrams**). Further reading Wilson (1979), Cardy (1993).

[38]Further reading Falicov (1992).

[39]Further reading Stein (1989), Huse *et al* (1992), Pobell (1993).

CHAPTER 14

INTERFACIAL MATTER

When two pure materials are in contact they may mix intimately to form a solution or homogeneous mixture. If they do not mix completely, the two phases share an interface that has properties that differ from those of either of the pure materials.

14.1 Windows

The region or domain at the interface between phases is of great practical importance in industry. It is also of considerable theoretical interest to scientists because it provides a kind of 'window' on bulk materials.

For most purposes, the term 'interface' is preferable to 'surface' because it conveys the fact that there are two phases involved. Except for solids in high vacuum, where there really is a 'surface', condensed materials are usually in contact with air, another gas or their own vapour. The term 'interface' also implies that this structure is not just an exposed 'surface' of the interior material. Indeed we can think in terms of an interfacial 'domain' or even interfacial 'phase' ('interphase'), with properties significantly different from those of the bulk phases. In particular, if there are surface-active molecules (surfactants) present in a solution, there is no doubt that the interfacial phase is very different from the bulk.

In pure materials, cohesive properties are important, and we ask to what extent the molecular particles experience attractive interactions. Similarly, at interfaces between **dissimilar** materials the **adhesive** properties are important. Cohesion parameters (Chapter 12) are just as useful for the characterisation of these heterogeneous systems (made up of more than one phase) as for homogeneous systems. At the molecular level, information on interactions at interfaces of solids and liquids is accessible from atomic force microscopy[1]†. We often use models based on 'interactions' or 'attraction' in talking about interfaces, although consideration of energy models is equally useful. Surface properties are particularly important in the case of microclusters where most atoms are on or near the surface.

14.2 Interfacial Energy

Interfacial energy is the excess energy[2] due to the existence of an interface. This excess energy arises from unbalanced molecular interactions associated with interfacial molecular particles compared to the more symmetrical interactions with neighbouring particles that occur in bulk materials. There are numerous practical examples of the tendency of materials to minimise their surface areas. A pool of molten metal is very smooth and flat, and glass manufacturers make high-quality window glass by floating molten glass on top of the molten metal. Similarly, one can make highly spherical particles by allowing drops of liquid to freeze as they fall. The oldest technology of this kind was lead shot made in a shot tower. Microgravity conditions would provide even more highly spherical particles.

In terms of interactions, molecules within a liquid experience cohesive interactions identical in all directions. However, those near interfaces have fewer neighbours closer to the interface and so experience attractive interactions towards the bulk (figure 14.1). In terms of an energy model, the molecules near the interfaces have more energy relative to those in the bulk, so interfaces are thermodynamically unfavourable. When a liquid or

† Endnotes for this chapter can be found on page 315.

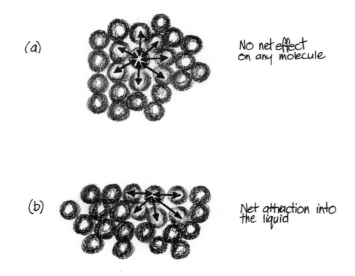

(a)

No net effect
on any molecule

(b)

Net attraction into
the liquid

Figure 14.1 Cohesive effects (a) in the interior of a liquid, and (b) at the surface.

solid is in equilibrium with its own vapour, we call this extra energy the surface free energy or 'surface energy'. Usually we apply this term also to the interfacial free energy of a condensed material in contact with air, which has a very similar value. (The older term for surface energy was 'surface tension'. There **is** a surface tension, but 'tension' is a force and not an energy.)

14.3 Liquids Can Flow Uphill

The most obvious question to ask about the interaction of a solid surface with a liquid is whether or not the liquid 'wets' the solid. We can contrast the 'beading' of water on a greasy plastic surface with the thin film that forms on a perfectly clean surface, especially in the presence of a detergent.

We can explore surface behaviour with an energy model to

compare the values of the various interfacial energies. In particular, we determine whether the new solid–liquid interfacial energy is less than the combined solid–air and liquid–air interfacial energies lost during the formation of the interface. A material with a lower 'surface energy' therefore wets one with a higher surface energy. The surface energies of metals are much greater than the surface energy of water, which in turn is greater than that of oil.

Surface scientists measure 'wettability' experimentally as the 'contact angle' of a liquid with a solid. The contact angle

- may be zero for complete wetting where there is a strong adhesion between the liquid and solid. Here the film of liquid lies 'flat' on the surface, like an aqueous detergent solution on glass,

- ranges from zero to ninety degrees for spreading systems, for beads of water on a greasy surface, for example,

- can be greater than ninety degrees when the liquid tends to 'shrink away' from the solid, like beads of pure mercury on clean glass.

Surprisingly, given the right conditions, a drop of water on a nonuniform, sloping surface can flow uphill. Treat a surface of silicon to provide a water-repellency gradient and tilt it so that the 'hydrophobic' ('water-fearing') end is lower than the 'hydrophilic' ('water-attracting') end. A drop of water placed at the lower end will then move slowly upwards[3]. A more common demonstration of the resultant force on a liquid that wets a solid is 'capillarity'. Here the liquid rises in a narrow tube until the gravitational force exerted by the liquid column balances it. The opposite effect occurs with a column of mercury in glass tubes, where the cohesive interactions within the mercury exceed the adhesive ones between mercury and glass, so the liquid level is depressed.

There is an interesting observation about the wetting of fibres. At first sight, liquids do not appear to wet some fibres, because the liquids form strings of droplets on the surface. However, closer observation shows that not only is the surface wetted, but there is a capillary flow of liquid from a small droplet to a neighbouring larger droplet. On the basis of the energy model, this minimises the surface area and therefore the surface energy, as one would expect. Also, we can use the rate of this flow to estimate the thickness of the liquid film[4].

Solids do not deform reversibly, so researchers cannot determine their surface energy values directly as they can for liquids. However, they can wet solids reversibly with liquids of known surface energy and so determine the solid surface information indirectly.

Surface energy can also drive a form of convection called 'thermocapillarity'. Small temperature variations at an interface cause this effect, and although obscured by gravity-driven thermal convection on Earth, it becomes apparent in the microgravity environment of space[5].

14.4 Adsorption

Because of the 'unsatisfied' molecular interactions at the surface of a material compared with those in the interior, in most natural situations that surface is 'contaminated' by 'adsorption' of other materials. If we want to study the original surface this is a great disadvantage, but there are practical situations where we can take advantage of adsorption.

Two terms that are often confused are 'adsorption' and 'absorption':

- Adsorption is the process where molecular particles (atoms or molecules) of one material attach themselves to the surface of another material.

- Absorption is the process where a gaseous or liquid material flows or diffuses into a porous solid (like a sponge) by way of pores or cracks.

This terminology is made even more confusing by the fact that the process of absorption into porous materials is often accompanied by adsorption onto the pore walls.

Adsorption is the result of interactions between two materials, and chemists define two limiting cases. At one extreme are the relatively weak van der Waals interactions resulting in short-term 'physical adsorption' or 'physisorption'. The extent of interaction can range all the way up to interactions which are as strong as 'normal' chemical bonds: long-term 'chemical adsorption' or 'chemisorption'. The activation model of chemical reaction kinetics is also applicable to the chemisorption process. On the other hand, physisorption is rapid and reversible, without the special energy requirements of 'activation'.

In the thermodynamic model, we express the extent to which adsorption stabilises the system in terms of the 'adsorption energy'. Small variations in the adsorption energy from system to system make very considerable differences in the typical 'adsorption time' or residence time on the surface. There is an **exponential**[6] relationship between adsorption energy and adsorption time, so relatively small differences in adsorption energy values cause very large differences in adsorption times. In physisorption, adsorption times are as short as a few hundredths of a nanosecond, while chemisorbed materials can remain on the surface for periods long on a human time scale. Physical diffusion of materials to the interface rather than the adsorption itself limits some of the most rapid adsorption processes.

14.5 Metal Surfaces

A clean cleavage surface of a crystal is different from one that has been heat treated or polished. In particular, there are fundamental differences between surfaces ground or abraded with a **harder** material, and those polished with a **softer** material such as jewellers' rouge or iron oxide. Low-energy electron diffraction (LEED) shows that grinding has little effect on the nature of the crystallinity. However, polishing leaves a relatively deep and nearly amorphous (glassy) surface layer that is able to flow into scratches. This softening, which we can observe whenever we polish brass or copper, is rather analogous to melting. In contrast, electropolishing (in which electrolysis with an alternating current repeatedly dissolves and replates metal) leaves a more nearly 'normal' surface crystal structure.'

A metal surface can act as a 'trap' for electrons, which are unable to pass an energy band gap into the bulk crystal. Physicists detect the resulting quantum mechanical electron standing waves by scanning tunnelling microscopy.

What I have said so far refers to the surface properties of pure metals. In pure metals, the environment of each of the bulk atoms is identical, but the interatomic interactions involving the surface atoms are different. There is now interest in the chemical and electronic processes occurring between the atoms of a bulk metal and a single layer of **different** metal atoms (a monolayer) supported on it. Such systems are of practical importance in catalysis and microelectronics[7]. Transfer of electron density towards the metal with the lowest valence band occupancy is important in these interactions between dissimilar metals. The larger the charge transfer the greater is the adhesive energy of the bimetallic bond.

An interesting application of the use of liquid metal surfaces is the proposal to construct lightweight and inexpensive astronomical telescopes from liquid mercury. The natural shape of the surface of a liquid rotating under gravitational force is a parabola, just what is required to focus parallel light to a point[8].

14.6 Thin Films

When dealing with the interfacial properties of liquids, we often encounter thin films of liquid. 'Bubbles' are a common form of vapour–liquid–vapour system. (Although we also call the cavities of vapour formed in liquids during boiling 'bubbles', strictly we should reserve this word for the thin-film bubbles.) Other phenomena involving thin films of liquids are wetting and capillarity, while clouds are very small liquid droplets suspended in a vapour. Nucleation events are often critical rate-determining steps in processes occurring in thin films and small dispersed particles.

In situations such as soap bubbles, we can observe thin layers of water only when they are 'protected' by surface-active agents: molecules that accumulate on both surfaces of the water film and stabilise it.

14.7 Surfactants

There is a close relationship between solubility or phase separation on the one hand and surface phenomena on the other. A component soluble to a limited extent in a mixture may demonstrate surface activity by reducing the interfacial free energy of the solvent and accumulating in excess concentration at the interface. This leads to the stabilisation of a foam film at a liquid–air interface. It is also possible for an insoluble component of a mixture to reduce the interfacial free energy at a liquid–air interface by spreading spontaneously as an insoluble film at the interface. In this case it has a foam-inhibiting effect. So two very similar compounds with equal surface activities and both reducing interfacial free energy can differ greatly in their behaviour because of slightly differing degrees of bulk solubility. Surfactant molecules (surface-active agents, those that accumulate at surfaces) such as soaps and detergents contain long chains of carbon atoms that would not normally be soluble in water. However, when such a chain carries an electronegative atom like oxygen or another group that can carry an electric charge, this group can interact strongly with the polar water molecules. This interaction with the water can be sufficient to allow the whole molecule to dissolve. So surfactant molecules contain two chemical groups:

● One is hydrophilic ('water-loving', highly soluble in water) and

often called the polar head or hydrophilic head of the
molecule.

• The other is hydrophobic ('water-fearing', water-insoluble)
  and sometimes called the aliphatic tail.

These two parts, if they were not chemically bound together,
would be highly immiscible. Consequently, chemists call such a
surfactant molecule 'amphiphilic', meaning 'loving both sides'.

If the water also contains salts, the dissolved ions can interfere
with the solution process, inhibiting the solution of organic
materials. There is a well known 'salting-out' effect, where
chemists can precipitate apparently soluble surfactants from
aqueous solution when they add a salt such as sodium chloride
(common salt).

In very dilute aqueous solutions, surfactants behave more or less
as ordinary solutes, although the hydrophobic groups on the
surfactant molecules tend to accumulate at the surface of the
liquid. Here they form a fluid 'skin', with the hydrophilic portions
of the molecules interacting strongly with the water phase and so
ensuring that the molecules remain in solution, up to the
equilibrium solubility limit. The surfactant reduces the surface
energy, and stabilises the production of a foam. This layer may be
only one molecule thick (a monolayer), providing a simple
technique to determine the cross-sectional area of molecules.
Surface scientists can use the same technique (known as a 'film
balance') on a smaller scale and in a more precise manner in the
laboratory. They can also observe through an optical microscope
the fluorescence from small quantities of certain fluorescent
surfactants incorporated as 'impurities' into an air/water
surfactant monolayer. Using polarised light and viewing at the
correct angle they can, even without fluorescers, detect those parts
of the surface bearing a monolayer: they can detect with a
microscope a single layer of surfactant molecules[9].

If we exceed the solubility limit, the surfactant molecules
associate in other ways. One of these is as dynamic 'micelles',
groups of surfactant molecules gathering as tiny globules within

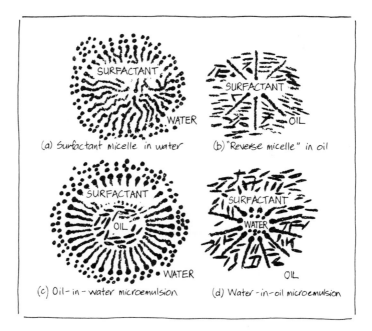

Figure 14.2 Structures involving surfactant molecules. (a) A surfactant micelle in water, with hydrophilic surfactant 'heads' in the water and hydrophobic 'tails' within the micelle. (b) 'Reverse micelles' of surfactants in oil, with the hydrophobic tails of the surfactant molecules in the oil and hydrophilic heads within the micelle. (c) Oil-in-water microemulsion stabilised by surfactant molecules, with hydrophobic tails in the oil and hydrophilic heads in the water. (d) Water-in-oil microemulsion stabilised by surfactant. After Rees and Robinson (1991).

the aqueous phase (figure 14.2). As micelles, the hydrophobic parts of the surfactants minimise their interaction with the water molecules by clustering with their 'tails' together as small colloidal aggregates. Their hydrophilic 'heads' point outwards into the aqueous medium. They are like tiny oil drops surrounded by a hydrophilic layer, but they are in solution rather than forming a separate phase. (In oil solution the orientation of the surfactant molecules is reversed. The hydrophilic heads cluster together within the micelle while the hydrophobic tails on the outside interact with the oil phase.) The particular surfactant concentration at which micelles begin to form is the 'critical micelle concentration'. The interiors of the micelles provide a local environment that is very different from the one outside, and these sometimes allow reactions to proceed much more readily than would otherwise be the case. In some ways and to a limited extent these micelles mimic enzymes. Micelles typically have diameters of the order of nanometres, with some of the properties of colloids. In other situations, some are 'wormlike' weakly bound structures classified as 'superpolymers' that enhance the viscosity in a manner similar to that of polymers[10]. Systems in which the particles are smaller than the wavelength of visible light have a high degree of transparency.

Where an oil–water two-phase situation exists, surfactant molecules gather at the oil–water interface, with a resulting reduction in the interfacial energy. This facilitates the formation of an emulsion, and in the right conditions there is a microemulsion. In terms of the thermodynamic model this is a stable single-phase system because of the entropy advantage gained from oil–water–surfactant interfaces, which are of fluctuating dimensions and shapes. In microemulsions the amount of surfactant controls the droplet size, and interfacial areas are extremely high as well as being rapidly changing.

The structures associated with surfactants at interfaces are reminiscent of liquid crystals[11], with domains in which there is ordering in one direction but freedom to flow in others. However, because surfactants are flexible long chain molecules, they differ from the rigid-rod liquid crystal phases.

14.8 Bubbles

Gas cavities rising through liquids (such as those in a glass of freshly poured beer or mineral water) form bubbles as they move through the interface of the liquid with air. In the absence of surfactants (as in mineral water) the bubbles formed at the surface burst almost immediately, but if surfactants are present (as in beer) they last longer.

Whether we are simply blowing a bubble from soapy water, or carrying out a precise laboratory film balance experiment, we find that a film of water containing surfactants acts very much like an elastic skin. This 'elastic skin' is a very good model to use for the interfacial layer. The surfactant layer is protecting the aqueous film. The 'elastic film' effect depends on the presence of the surfactant molecules and is absent in pure liquids. This is 'Gibbs elasticity', following its description by J Willard Gibbs. The stretching of the film of liquid, although increasing the surface available for surfactants to occupy, decreases the number of surfactant molecules per unit area, so the surface energy increases. Surfactant molecules take time to diffuse to the newly stretched surface (the 'Marangoni effect'). Because their concentration per unit area is therefore less than it would be at equilibrium, so the surface energy is even greater than at equilibrium. These effects describe not only the 'elasticity' but also the stability to fluctuations of films containing surfactants. Any attempt we make to increase the film area results in an increase in surface energy, which opposes the expansion. They are properties not demonstrated by pure liquids, which suggests why it is easier to blow 'soap' bubbles than to blow 'water bubbles'[12].

14.9 Foams

Foams are agglomerations of bubbles formed from a gas (often air) contained within films of liquid (usually water-based). They are metastable (only temporarily stable) structures formed during agitation of a liquid containing a surfactant. We can discuss the balance between effects tending to maintain the foam and to collapse it in terms of the energy model, and in this case the interfacial energy is of particular importance. The interface is not as energetically unfavourable for surfactant molecules as it is for water molecules, so the interfacial energy is lower than in the case of pure water.

The stability of foams has received attention for both theoretical and practical reasons. The 'surface tension' acting inwards in a bubble balances the air pressure inside. It is instructive to ask what would happen if we had two identical soap bubbles at each end of a connecting tube equalising the air pressure. In the simplest model, this situation is inherently unstable. We would expect the smallest fluctuation of size making one smaller than the other to lead to the internal pressure of the smaller bubble increasing so that even more air flowed to the larger. This would lead rapidly to the destruction of the smaller bubble. However, in practice the stretching of the larger bubble is opposed by its tendency to contract (the Gibbs elasticity) and by the delay in surfactant molecules diffusing to a newly formed surface (the Marangoni effect). The result is that infinitesimal fluctuations in the foam bubbles are stabilised.

When first formed, 'wet' foams consist of bubbles, each bubble having a wall thick enough to maintain an approximately spherical shape (minimising the surface area for a particular volume). As the water drains away, the liquid films become thinner and the bubbles interact with each other. This 'dry' foam is an array of polyhedral (many-sided) bubbles, rather than spherical bubbles.

Foams with known and reproducible characteristics have applications for[13]

- distributing relatively small amounts of material uniformly over a wide area (like hair-setting foam and cleaning foams), and for

- isolating and trapping relatively small amounts of material within a small area (like froth flotation for ore separation).

14.10 Aerogels

Imagine a gel that remains the same size when all the water is replaced by air. It is an 'aerogel': an extremely low-density material with a porosity as high as 98%.

Usually, when the liquid evaporates from the surface of a gel, the tension within the interface creates a concave meniscus in each of the pores of the gel. This results in a compressive force on the gel as a whole that causes it to collapse. S S Kistler of Stanford University in the early 1930s removed the water from a gel of sodium silicate ('water glass') in an autoclave or pressure vessel under supercritical high-temperature, high-pressure conditions[14]. In supercritical fluids there is no distinction between liquid and vapour and therefore no meniscus and no surface energy. The process therefore removed the water slowly, leaving an undamaged porous gel of the original size and shape, which he called an 'aerogel'. The process was slow, and we now can use other techniques, but the principle remains the same.

Scientists investigate the nature of the porosity of aerogels experimentally. Because of Rayleigh scattering, aerogel silica appears yellowish when looked at with a light source behind it, but bluish when illuminated from the front and viewed against a dark background. These observations and more precise measurements indicate pores roughly 50 nanometres in diameter. (For comparison, recall that a typical atomic diameter is a tenth of a nanometre). The gel structure determines percolation and insulation properties. In terms of the pore structure, only below certain critical densities is there a coherent path through the pores, permitting the direct movement of fluid through a sample. More importantly for its thermal insulating properties, only **above** certain critical densities is there a continuous silica structure, which provides a continuous thermally conducting path through the solid material. Therefore low-density aerogels are excellent thermal insulators. Associated with these complementary observations is the fact that aerogel structures are fractals[15]. The percolation thresholds are modelled by computer simulation, using random filling of sites in a lattice pattern.

Such silica aerogels, which combine transparency to visible radiation with valuable thermal insulating properties as well as low acoustic impedance, have potential practical applications. One of these already in use results from its refractive index. In this case the waves being studied are high-energy charged protons and other particles that emit radiation ('Cerenkov radiation') when moving in a medium with velocity greater than the speed of light in that medium. It turns out that the refractive index of aerogels (intermediate between those of gases and liquids) is just right for the construction of Cerenkov detectors. More mundane but just as important are applications of aerogels in house insulation and solar energy collection[16].

14.11 Sticky Solids

The attractive or cohesive properties occurring between molecular particles are also important in the physical interactions that occur between macroscopic solid bodies and between solid bodies and bulk liquids[17].

Molecular particles (which may be atoms or molecules or ions) in condensed materials stick together because of the attractive interactions characteristic of condensed matter. As I have shown, this causes cohesion between similar materials and adhesion between different ones. When drops of liquids touch, they coalesce. Many solids adsorb gas molecules strongly. Why, then, don't macroscopic solid objects usually stick together?

Situations where macroscopic solids do stick together seem to have special requirements:

- Cements, plasters, pigments, the very small rubber particles in emulsion paints on walls, and cellulose fibres in paper remain bonded together through a variety of interactions, including hydrogen bonding.

- High-temperature bonding of clay and bone particles produce pottery and porcelain.

- Opals are composed of tiny silica spheres that have been in close contact at high temperatures and pressures for long periods of time.

- Mollusc shells contain minute crystals of calcium carbonate bonded with polymers.

It turns out that solid particles stick together only if there is intimate contact between them. The attractive intermolecular van der Waals interactions are relatively short range so a single layer of adsorbed atoms or the slightest irregularity in the surfaces can reduce the attraction to a negligible level. This is in contrast to electromagnetic and gravitational interactions which are of longer range: a magnet sticks to steel even if the surface is painted. So the two factors that limit solids being observed sticking together are:

- surface contact area (determined by object size and roughness), and
- surface contamination.

Rubber surfaces exhibit 'stickiness' far more readily than do glasses or crystals. Rubbers 'give' to conform to the shape of the other surface and rubbery materials are less likely to adsorb water (like silicate glasses) or oxygen (like many metals). In papermaking the presence of water as a 'plasticiser' for cellulose fibres allows them to come into intimate contact and establish hydrogen-bonds and other van der Waals interactions that persist after the paper dries.

Sensitive measurements on the interactions between clean, convex solid spherical surfaces like glass lenses show that they attract each other. Although completely 'dry' they even appear to 'wet' each other, just as if there were liquid present: there is perfect molecular 'meeting', comparable to that occurring when a liquid wets a solid. The process of hardening of emulsion paints involves the evaporation of water so that adjacent tiny rubber spheres touch each other and pull each other together strongly enough to eliminate pores. This mutual attraction of solid elastic bodies is not the same as the coalescing of liquid drops but is due to a combination of van der Waals interactions and geometric factors. The surface-to-volume ratio is much greater for small particles. Also, the size of the contact spot when two spheres touch is much closer to the size of the spheres when the spheres are small (like the rubber particles in latex paints) than when they are as large (like squash balls). For these reasons we do not notice the attraction between large objects.

The addition of contaminating liquid in the vicinity of what would otherwise be good adhesion or cohesion between solids decreases the effect. It is only if the surfaces are rough that liquids can increase the stickiness by filling what would otherwise be spaces between the solids[18].

14.12 Sintering

Some solids possess sufficient plasticity to flow slowly at temperatures below their melting points. Usually the effects become noticeable at temperatures of between one-half and two-thirds of the melting point on the absolute temperature (kelvin) scale. Both bulk and surface diffusion become appreciable, as demonstrated by:

- the 'filling' of surface scratches on silver when it is maintained at a temperature just below its melting point,

- ice cubes 'fusing' together even at minus ten degrees Celsius.

The surface attraction of small particles of metals or of metal oxides can effectively pull them into contact and cause them to permanently fuse at temperatures well below their melting points. This effect supports the practical technologies of powder metallurgy and ceramic manufacture. Attempts to model systems in order to predict areas of contact and rates of flow have ranged all the way from treating them as viscous drops of liquids to assuming that they are elastic spheres. Success has been limited[19].

14.13 Aggregates

Powders can agglomerate even at low temperatures under the right conditions (intimate contacts of clean surfaces). It is now believed that the solid bodies in the solar system are the result of the aggregation of solid particles, with the ability of the particles to adhere being significant in the resulting growth. An everyday example of these randomly grown, fractal clusters is the 'tenuous' aggregation of soot and ash particles. Another example is the growth of colloids in colloidal suspensions[20]. Structures arising from the growth of a cluster on a 'seed' particle by successive random movements of other particles involve a 'random walk' diffusion-limited aggregation process. They exhibit a fractal geometry in which a degree of order arises as a result of random events[21]. Very small particles may also be so 'sticky' (attract each other so strongly) that they cannot organise themselves into close-packed structures by 'exploring' various configurations. The resulting effective densities are low. In an extreme case, very small sand particles of the order of twenty nanometres in diameter when poured into a container produce a material one-twentieth the density of water. Computer simulation can use the 'ballistic aggregation' model, each new particle added to a growing network sticking irreversibly when it hits an existing particle. If some repulsive interaction is added to the model so that the collisions are less 'sticky', incoming particles are free to seek lower-energy sites with more near neighbours and a resulting increase in density[22].

Pharmaceutical manufacturers utilise the 'stickiness' of powders even when 'dry' during the process of making tablets. If small solid particles cannot crack under compression, they flow and fuse together at points of contact. In **granulation**, a powder is stirred in the presence of a liquid, which then subsequently evaporates. Most **filtration** does not rely on particles being trapped by lodging in the holes. Rather, while passing relatively freely through channels of greater diameter, the particles stick to the channel walls.

14.14 Nanotribology

Tolerances in the construction of engineering components now approach nanometre dimensions, so it is necessary to observe and model interfaces and lubrication on an atomic scale. 'Nanotribology' or 'molecular tribology' is the study of friction (from the Greek word *tribos* meaning 'to rub') on an atomic or 'nanometre' scale[23]. An atomic force microscope shows that at separations of the order of nanometres atoms are transferred from a gold metal surface to a nickel probe above it. Even on this atomic scale, gold with the lower surface energy tends to 'wet' the higher-energy nickel surface, resulting in an intermetallic junction or adhesion. The result on a macroscopic scale is friction, or resistance to shear. Atomic force microscopy can provide information on both normal (up-and-down) and lateral (frictional) forces as the probe tip moves across a surface.

A 'lubricant' takes the form of interlocking branched-chain hydrocarbons in molecular-scale layers between the metal surfaces. This acts as an atomic-scale barrier to the formation of metal–metal interactions that cause friction. Observations show that materials do not act as lubricants just because they are viscous but because they present an effective barrier to atom-scale welding[24].

14.15 Adhesives

All the factors relevant to cohesion in a pure material are just as relevant to the process of adhesion, the process of making two different things stick together. In addition, it is necessary that an adhesive should wet the surface. As pointed out by Ben Selinger[25] the reason that a baked cake adheres to its baking tin is that the water in the cake mixture wets the metal, then leaves behind an adhesive deposit during the cooking process. Greasing the tin to prevent the cake sticking has the effect of preventing it being wetted by the aqueous material. Even water applied as a liquid and frozen makes a reasonably good adhesive below zero degrees Celsius!

When polymer cements (concentrated solutions of polymers in organic liquids) are used to join plastic components, the solvent penetrates and swells the plastic. This frees the polymer chains in each part so that they can become entangled with each other, and evaporation of the solvent leaves the parts permanently joined.

One of the misleading models some of us have in our minds is that the force necessary to break an adhesive joint is always proportional to the area in contact. Rather, when it comes to peeling an adhesive film, the stress concentrates in a very small region near the peel line. The failure process is similar to crack propagation and best described in terms of surface energy[26].

Endnotes

[1]See **8.10 Surface Images** and **14.14 Nanotribology**.

[2]Strictly, free energy: see **5.5 Death and Taxes**.

[3]Further reading Chaudhury and Whitesides (1992).

[4]Further reading de Gennes (1992).

[5]See **2.7 Microgravity**. Further reading Amato (1992d).

[6]See **2.2 Perspective**.

[7]Further reading Rodriguez and Goodman (1992). See also **12.9 Catalysts and Enzymes** and Friend (1993).

[8]Further reading Borra (1994).

[9]Further reading Barnes (1992).

[10]'Nanoemulsions' are described in **5.2 Thermodynamic Model**. See also **12.9 Catalysts and Enzymes** and **15.4 Flowing Polymers**. Further reading Lubensky and Pincus (1984), Rees and Robinson (1991).

[11]See **15.8 Liquid Crystals**.

[12]Further reading Aubert *et al* (1986).

[13]Further reading Aubert *et al* (1986).

[14]See **12.4 Phase Diagrams**.

[15]See **3.8 Fractals**.

[16]Further reading Fricke (1988).

[17]Further reading van Oss *et al* (1988). The term 'sticky solids' is the title of an article by Kendall (1980).

[18]Other factors associated with sticking (and unsticking) are discussed in **14.15 Adhesives**. Further reading Kendall (1980).

[19]Further reading Kendall (1980).

[20]See **15.6 Colloids**.

[21]See **3.8 Fractals** and **2.4 Condition Critical**.

[22]Further reading Stewart (1992), Donn (1990), Lubensky and Pincus (1984, p 50), van Oss *et al* (1988), Kendall (1980), Healy (1992).

[23]Further reading Belak (1993a, b), Salmeron (1993), Overney and Meyer (1993), Landman and Luedtke (1993), Robbins *et al* (1993), Harrison *et al* (1993).

[24]Further reading Yam (1991).

[25]Selinger (1989).

[26]Further reading Kendall (1980, p 284), Barton (1982).

CHAPTER 15

SOFT MATTER

We are now ready to look at materials with structures even more complex than those of mixtures and solutions and interfaces. They are materials that include the components of living things.

15.1 Soft Matter

The term 'soft matter' is now being used by some scientists as a simple descriptive title for a broad and diverse range of noncrystalline materials. They include polymers, surfactants, liquid crystals, colloids, emulsions, surfactant films and other materials with properties more complex than newtonian liquids because they possess additional 'structure'. This general descriptive title is perhaps preferable to the approximately equivalent terms 'complex fluid' or 'structured fluid'. We can use the expression to describe materials as diverse as syrup, egg-white, sauces, paint and toothpaste simply because it is a word-model that accommodates a greater range of materials. In a similar way, some scientists use the name 'ultraweak solid' for some of the soft matter structures with the weakest interactions, like colloids and emulsions[1]†.

15.2 Polymers

These materials arise as a result of an assembly of the long and usually flexible 'macromolecules' or 'polymer' molecules. It is

† Endnotes for this chapter can be found on page 340.

helpful at the outset to distinguish between the terms 'plastic' and 'polymer'. Originally 'plastic' was an adjective meaning 'readily moulded', but now we also use it as a noun referring to mixtures of polymers with additives that have 'plastic' properties. Mechanical stress at room temperature causes many pure polymers to fail by brittle fracture rather than by stretching (plastic flow). However, polymers remain soft and pliable when polymer technologists incorporate in them other compounds called 'plasticisers'[2]. **Polymers** are pure chemical compounds, although they do have a distribution of molecular lengths arising from different numbers of segments (macromolecules of various chain lengths). Commercial **plastics** are not usually pure chemical compounds, but are mixtures incorporating polymers, plasticisers, antioxidants and other additives, formulated for various applications.

Even without additives, polymers show a very wide range of physical properties as a result of the various types of chemical bond joining the segments. Some polymers such as 'polyimides' containing a rigid bond structure are stable to over three hundred degrees Celsius, while others soften and flow at low temperatures. Some are ordered and crystalline at lower temperatures. Others retain a disordered amorphous or glassy structure rather than a crystalline nature at low temperatures. Most form a liquid phase above a certain temperature, but these may be liquid crystals rather than 'normal' liquids.

One of the usual classifications of polymers is into 'linear polymers' and 'crosslinked polymers'. Linear polymers (those with long but separate macromolecules) form materials that are 'thermoplastic' like polyethylene. They soften and eventually melt with increasing temperature. On the other hand, crosslinked polymers, with an 'infinite' three-dimensional structure, are 'thermosetting'. Like the first wholly synthetic plastic ('Bakelite') they do not melt as the temperature increases, but remain solid and eventually decompose if the heating continues.

A flexible polymer acquires greater rigidity and toughness if there is crosslinking between polymer chains. Indians of the Amazon carried out polymer crosslinking by oxygen. They coated their feet in the sap of the hevea or rubber tree and allowed oxygen from the air to convert it into rubber boots. In a more up-to-date application, vulcanisation of rubber in vehicle tyres involves crosslinking by sulphur. The tanning of leather involves the crosslinking of 'collagen' (a protein macromolecule) in skins.

Most polymers occur in the amorphous (disordered, non-crystalline) state. Some of these are brittle glasses at room temperature while others are above their glass transition temperatures and are 'rubbery'[3]. In appropriate circumstances some polymers will crystallise, the densities of the crystalline forms being between five and twenty per cent greater than densities of the amorphous states. Heating a thermoplastic polymer eventually generates a liquid (a 'polymer melt'). Liquid polymers have a structure rather like a can of spaghetti in tomato sauce. A better model is a can of worms, because the macromolecules are in constant movement. Each polymer chain is able to move only with difficulty (as a worm or snake would) through an array of obstacles made up of other macromolecules. Polymer scientists call this type of movement 'reptation', from the Latin word meaning 'creep'[4].

We can use the reptation model to express the relationships between the movements of the whole molecule and the movements of the molecular segments. Important for the former processes are the rather long times (or low frequencies) characteristic of movement of the macromolecules as a whole, and the number of segments in the chain. In contrast, there are the very short-time (or high-frequency) responses characteristic of movements of the individual segments. This dual structure creates 'viscoelastic' properties, a combination of viscosity and elasticity[5], dominating the behaviour of polymer melts and solutions. Interesting examples of the application of the reptation time model are the interdiffusion and welding of polymer blocks held in close contact just above the glass transition temperature for various periods. These models show that polymer molecules can 'choose' various types of interaction (segment–segment, segment–solvent or segment–surface) and can 'explore' a wide range of configurations, making them significantly different from monomer materials. We can also use reptation models in designing materials as two-dimensional 'obstacle courses' for the fractionation of large molecules[6].

Let us return to the spaghetti model of polymers and look at the properties of the crystalline or glassy materials. There is currently considerable interest in 'uncooking the spaghetti': straightening out and orienting the jumbled coils and putting them back together as they were in their original dry state. Polymeric materials when oriented in this way have the potential to have many times the strength and stiffness of steel on a weight-for-weight basis, and some are also highly conducting. The first commercial product (with strength in the direction of alignment, but not perpendicular to it) was Kevlar®, the du Pont 'aramid'. (An aramid is a polymer based on aromatic amides.) One can make oriented polymers in at least three ways[7]:

- by organising them as liquid crystals, or

- by diluting the chains and disentangling them before removing the solvent, or

- by polymerising monomers that have already been lined up within an oriented solid.

15.3 Polymer Solvents

We can consider polymer solutions, as we can consider all equilibrium systems, in terms of the thermodynamic model[8]. The entropy state of a polymer solution is more positive than that of the amorphous polymeric state because the molecules in solution are more free to move and rotate. However, the entropy increase during dissolution is less than that occurring when two monomeric liquids (with molecules of approximately equal size) mix, so polymers are soluble in liquids only if there are more energetically favourable interactions between liquid and polymer than between liquid and liquid, and between polymer and polymer. Even when a particular polymer–liquid pair shows complete miscibility at high polymer concentrations, it is not unusual for two phases to form when we dilute with further liquid. This is very inconvenient if it occurs when we add more liquid as a solvent to try to clean our brushes after painting!

As an example of the kind of situation that occurs, consider those polymers which are based on a carbon skeleton and which have most of the substituent hydrogen atoms replaced with fluorine atoms to form 'fluoropolymers'. These widely used protective coatings and sealing materials are insoluble in most organic liquids. Until recently, industry used liquids known as chlorofluorocarbons (CFCs) for both the synthesis and the processing of fluoropolymers, but CFCs contribute to the depletion of the ozone layer. It was therefore necessary to identify alternative solvents. Rather than testing all the thousands of pure liquids and the very much greater number of possible combinations of liquids on a trial-and-error basis, it is clearly preferable to use models. These models help us predict which properties of liquids are most relevant in generating favourable interactions with (in this case) fluoropolymers.

It is possible to vary the nature of these interactions. We can do this not only by using different liquids, but also by altering the density or closeness of packing of the molecules by means of pressure changes. While this requires very high pressures in the case of a liquid, it is much easier when we use a supercritical fluid. Because supercritical fluids are very compressible, we can obtain a wide range of densities and resulting solvent properties. Supercritical carbon dioxide is already proving to be a particularly valuable material as a solvent and extraction agent[9].

15.4 Flowing Polymers

As seen in discussions of viscous flow, there are situations where the flow of any material may be non-newtonian. In other words, the flow does not conform to the simple definition of the viscosity coefficient that holds in simple liquids, where the flow properties are independent of the rate of shear. However, flow properties of soft matter like polymeric fluids (melts or solutions containing macromolecules) are **always** observed to be non-newtonian. In fact, it is common to classify a fluid as either 'newtonian' or 'polymeric' depending on its flow behaviour. Simple observation of flows of polymer melts and polymer

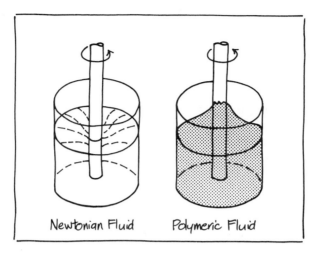

Newtonian Fluid   Polymeric Fluid

Figure 15.1 Simple observation of flows of polymers (and polymer solutions) shows that they differ fundamentally from the flows of newtonian fluids, as demonstrated by their behaviour on stirring. (After Bird and Curtiss (1984).)

solutions shows that they differ fundamentally from the flows of newtonian fluids.

You have all seen the surface of a liquid depressed when a vertical stirrer is rotating rapidly within it. If you try the same thing with a polymeric system the liquid climbs **up** the rod (the 'Weissenberg effect') rather than being depressed (figure 15.1). The same thing happens if you stir the liquid with a rotating magnetic stirrer at the bottom of the container: there is an elevated vortex rather than a depressed one. Also, when you stop the flow of a polymeric fluid, the material recoils. There is a 'fading memory' effect, intermediate between that of a newtonian material and an elastic material:

- a newtonian fluid has no memory of previous events,

- an elastic material has a complete memory of previous events,

- a polymeric fluid has a partial memory of previous events.

Unlike newtonian flow, where the stream of liquid contracts as it issues from a tube, a polymeric flow from a tube swells as it emerges. We know that we cannot siphon a newtonian liquid if there is gap between the surface of the liquid and the siphon tube, but a polymeric liquid will siphon across a gap. Some of the other effects are less immediately obvious, but just as significant. For example, if we drop two spheres into a newtonian liquid, the second sphere tends to overtake the first, but in polymeric fluids the distance between the spheres tends to increase. All these effects do not indicate just minor differences between newtonian and polymeric liquids. They are fundamentally different (or 'qualitatively different' rather than 'quantitatively different')[10].

Another word used for the study of non-newtonian fluids is 'rheology' and the measurements made on them involve 'rheometry'. This is the non-newtonian equivalent of viscometry. The development of simplified conceptual and mathematical models of macromolecular systems (such as 'beads' for the polymer segments joined by elastic 'springs' and rigid 'rods' for chemical bonds) is an important aspect of rheology. There are also models set up for assessing the molecular components of mechanisms for transmitting forces across a notional plane moving with the average velocity of a flowing fluid. One can distinguish three effects:

- momentum transfer involving movement of solvent molecules and polymer segments across the plane,

- intramolecular tension in bonds within polymer molecules lying across the plane, and

- intermolecular solvent–solvent, solvent–segment and segment–segment interactions across the plane.

Models based on temporary 'slip-link' networks with slippage of bead–spring systems at the points of entanglement are also possible[11].

Without going into details of these models, it is useful to explore some of the ideas involved and develop a series of models to make the properties of polymeric fluids clearer. If we apply a force to a 'simple liquid' made up of molecules with low molecular mass, the system 'forgets' the distortion very rapidly compared with 'human' time scales. These times are of the order of successive intermolecular collisions. On the molecular dimension scale, the interactions between solvent molecules and the flexible polymer segments also have these short relaxation times, less than a **nanosecond**. In contrast, in concentrated polymer solutions or polymer melts the relaxation times for disentanglement can be of the order of **seconds**. Consequently the relaxation times for macroscopic flows are also very much longer, resulting in both increased viscosity and a build-up in distortion that can make the viscosity depend on shear rate. We can ask concerning polymeric fluids: what is the size of the typical structure, and how can we scale up from the molecular level to the macroscopic level?

The great **bulk** of the macromolecules is one of the factors affecting macroscopic polymeric fluid dynamics, but this alone does not introduce many of the observed viscoelastic properties. Two other factors are important.

One is that macromolecules are tenuous or insubstantial. Each polymer chain tends to spread out to influence a large volume of the solution or melt (and therefore influence many other macromolecules and solvent molecules) rather than to contract into a compact space. This results in enhanced viscosity. Also, because the directional correlation between polymer segments more than a few units apart is negligible, polymeric fluids have tenuous spatial properties related to fractals.

(An exception to the existence of these tenuous properties arises in solutions of polymers in liquids that are poor solvents. Near the limits of solubility, where the polymer segments interact more strongly with each other than with the solvent, the preferred configuration is with the macromolecules coiled up into balls. As a result, the viscosity is lower.)

The other factor is the flexibility of macromolecules or polymer chains. They can adopt an effectively infinite number of configurations rather than being locked into one configuration. In the thermodynamic model, this randomness in macromolecular configurations is a reserve of entropy not present in simple liquids.

Alternatively, in terms of 'beads and springs' models, we can look at this property as the ability to store mechanical energy. Over short times polymer chains respond elastically, 'remembering' their previous configurations. After longer times the chains disentangle and 'forget' them[12].

Even more complex and interesting are 'electrorheological fluids'. These are slurries that transform in an electric field into gel-like states containing chains or fibrils of the suspended particles. An example is the structure of columns of starch macromolecules in oil that form in high electric fields. Visual effects designers used this type of material, neither a solid nor a liquid but yet another state of matter, as the liquid metal skin of the self-healing robot in *Terminator 2: Judgement Day*[13].

One usually thinks of polymers as being organic materials. However, there are also inorganic polymers, and as an example the viscosities of complex silicate melts such as those in the Earth's magma and volcanic lava flows are important[14].

15.5 Gels

In some cases the interactions between polymer chains in solutions are so extensive that the bonded polymer 'molecules' extend throughout the material. Instead of a polymer solution that can flow there is a gel with some of the properties of a solid. Mechanical perturbations can move from one side of a sample to the other. For example, dessert 'jelly' is largely water, with about 3% of the animal protein, gelatin. Without the water, the polymer molecules would collapse into a much smaller volume, and without the polymer network the water would flow[15].

The bonding within a gel may be chemical covalent bonds, van der Waals interactions or hydrogen bonds. It may involve the complex interaction of proteins, as in gelatin gels and some biological materials. Gel properties depend strongly on the nature and amount of solvent present, with a balance being reached between polymer–polymer and polymer–liquid interactions, as well as on the extent of crosslinking between polymer chains. Gels are 'viscoelastic', with complex flow properties[16]. In some systems the gel density reaches a state of criticality as one changes the composition, electric field or temperature. This provides the opportunity for us to design 'intelligent gels' as valves or pumps or artificial muscles: soft chemomechanical systems[17].

15.6 Colloids

Colloid systems constitute yet another 'state of matter'. Colloidal particles are nanoscopic, and colloid systems are suspensions of particles having diameters of the order of one to one hundred nanometres that do not settle out on standing. Michael Faraday has been described as an early 'nanotechnologist' because he studied metal colloids, like the ruby-coloured colloidal gold, in the mid-nineteenth century. Some of the actual samples he prepared survived unchanged today, the colloidal particles still in suspension. A twenty nanometre gold particle contains sixty thousand gold atoms, and the colloidal material has a very large surface area (about sixty square metres per gram) compared with that of macroscopic gold particles[18].

These colloid systems of heavy metals like gold are at the lower nanoscopic end of the particle size range. At the upper size limit of colloidal behaviour there are colloidal dispersions made up of particles that are much larger, but much less dense. Dispersed colloid systems of these bulky particles suspended in a liquid are further examples of 'soft matter'. Apart from their size, the colloid particles do not have the other properties that characterise polymer fluid dynamics. They are neither tenuous nor flexible, and so lack the elastic properties of polymeric fluids. (For this reason they make useful large-scale models of the ways in which molecular particles behave, because they are easier to 'see' than small molecules.) Examples are black ink (with colloidal carbon), xerographic toner (with colloidal particles of carbon black and polymer) and 'ferrofluids', discussed below. Bulky colloidal particles can have significant values of interaction energy (many times greater than thermal energy). Therefore they form aggregates by sticking together at the first contact, resulting in tenuous fractals. These 'aggregated colloids' show not only the properties characteristic of dispersed colloids, but also some of the increased viscosity effects characteristic of polymeric systems. These effects result from the tenuous structures and occur in food products such as ice cream and sauces[19].

The colloidal particles in the systems described so far exist in a random or disordered arrangement, just as the molecular particles in the liquid are random or disordered. However, the relatively long-range electrostatic interactions (called electrical 'double layers') that occur between some ionic colloidal particles can cause the particles to adopt uniform, ordered arrangements. If the spacings of the colloidal particles are comparable to the wavelengths of visible light they act as diffraction gratings and optical colour effects result. We can visualise this as a long-range ordered array of colloidal particles with repeat distances of the order of a hundred nanometres, supported by an array of liquid molecules with dimensions of the order of one nanometre and no long-range order. These systems are sometimes called 'colloid crystals' (analogous to 'liquid crystals'), the word 'crystal' referring to the long-range, hundred-nanometre-scale ordering.

The ionic colloid particles themseleves have 'true' crystalline ordering on a sub-nanometre scale, but this is not the reason for the term 'colloid crystal'. Here we are using the word 'crystal' for its mental image of long-range order, rather than for its more specific description of an array of close-packed ions or atoms forming a hard, dense material. In contrast, earlier we saw that the term 'crystal' is applied to certain glassy (noncrystalline) materials based on lead silicate because of some physical resemblance to 'real' crystals: hard, transparent and highly refracting. In maintaining the appropriate 'state of mind' to recognise a different 'state of matter' we must accept that language can be a barrier as well as a bridge to new concepts.

15.7 Superpolymers

A ferrofluid is a colloidal suspension in a liquid, such as oil, of very fine magnetic particles, such as ten nanometre cobalt particles (often coated with polymer or surfactant to prevent the formation of aggregates). The particle size is sufficiently small that each particle has only a single magnetic domain. These systems respond in unusual ways to applied magnetic fields because the magnetic energy is of comparable magnitude to the surface energy. An example is the formation of weakly bound chain-like structures we classify (along with the worm-like surfactant micelles) as 'superpolymers'. They occur in applications like printing inks and also in seals, suspended between magnetised bushes of bearings[20].

15.8 Liquid Crystals

Friedrich Reinitzer and Otto Lehmann identified liquid crystals as a different 'state of matter' towards the end of the nineteenth century. They noticed that cholesteryl benzoate on heating first melts to a cloudy liquid and then in a second stage changes to a clear liquid. Also, they saw bright colours when they examined the cloudy liquid with a polarising microscope as a result of its birefringence, its direction-dependent refractive index. Crystals, with definite structures, frequently show this birefringence effect, but liquids that are isotropic on a macroscopic scale do not. It was Lehmann who decided that they could be called 'liquid crystals'. (As discussed above for 'colloid crystals', the word 'crystal' is being used here for the image it conveys of long-range repetition or order, and not to imply some of the properties we often associate with the term.)

This type of material is not at all rare. Biological materials (such as DNA itself) tend to form liquid crystal phases, and you can see the birefringent or doubly refracting nature of striated muscle fibre in any butcher's shop[21]. Liquid crystals are a form of condensed matter in which there is no long-range positional order in the arrangement of the molecules, but they are orientationally anisotropic. There is long-range bond-orientational order or alignment associated with the long, rod-like organic molecules (unlike spherical molecules, which can have positional order but not orientational order (figure 15.2)). When one heats a crystal of some materials made up of long-rod molecules, the crystal loses

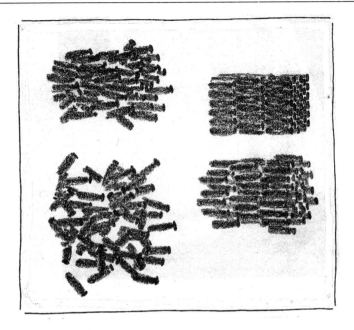

Figure 15.2 Clockwise from the top right are representations of an ordered solid crystal, a smectic liquid crystal with considerable layered order, a completely disordered liquid and a nematic liquid crystal with some orientational order. (After Templer and Attard (1991).)

positional order and becomes a fluid. However, it may retain the orientational order and form a nematic ('thread-like') liquid crystal state. At higher temperatures it loses this order also, resulting in a 'normal' liquid. If the rod-like molecules have the property of chirality or 'handedness' they may form a 'chiral nematic' phase or 'cholesteric' phase. These materials, named 'cholesteric' liquid crystals because cholesteryl benzoate was the first example observed, have a helical structure resulting from the 'rotation' imposed by each molecule on its neighbour. If the repeat length of the patterns in nematic phases is of the same order as the wavelength of visible light, this liquid crystal phase can act as a diffraction grating and generate visible colours. Because the spacing changes with temperature the material changes colour with temperature (it is 'thermochromic'), and numerous applications are now being developed[22].

In some liquid crystal phases the molecules line up in parallel configurations in successive stacked layers. Chemists describe this structure as 'smectic', from the Greek word for soap, because the mixture of soap and water accumulating in the bottom of most bathroom soap dishes is a smectic liquid crystal. Its complex stratified structure has prompted the use of several models in its description and analysis. Previously, there had been theoretical modelling[23] involving melting in a two-dimensional solid. We can study two-dimensional melting experimentally by drawing a free-standing film of a smectic liquid crystal material across a hole with

a wiper. We can form samples as thin as two molecular layers and examine them by in-plane x-ray scattering.

Scientists have also interpreted the planes of the smectic liquid crystals as successive stages of a one-dimensional, sinusoidal density wave within a three-dimensional fluid. (This may be a difficult model to visualise, but before the development of such models, liquid crystal materials were even more mysterious.) 'Chiral smectic' phases also occur, and if the molecules in such phases have dipole moments, then 'ferroelectric' (or 'helielectric') materials may form[24].

Other types of liquid crystal structure occur with surface-active agents. Here there is a partial ordering as a result of the hydrophilic or water-loving 'heads' and the hydrophobic or water-fearing 'tails' of the molecules optimising their configuration in an aqueous environment. These structures include sheets of 'bilayers', in which all the 'heads' of the molecules face the aqueous environment on the outer surfaces of a sheet. The hydrophobic 'tails' make up the interior of the sheet. The bilayer itself acts like a liquid, with the surfactant molecules able to move around in two dimensions but not able to leave the sheet. These bilayer sheets can also rearrange themselves into globular sacs, forming a simple 'cell' that models biological cells. Living cells do indeed have bilayer membrane walls, and liquid crystal cells may have been protocells preceding 'life' ('prebiotic' structures) on Earth. Other, more complex, bilayer structures exist in both chemical and biological systems. One of these is the red blood cell, which has no energy barrier or 'surface tension' associated with its interface with the fluid outside. This results in major fluctuations or 'flicker' in the cell shapes that are detectable with phase contrast microscopy[25].

One can easily alter the 'director' of a liquid crystal (the direction of alignment) by applying an electric or magnetic field. This changes the refractive index as well as the light-scattering and other optical properties, so combined with microprocessors liquid crystal displays have become part of our way of life. Liquid crystal ferroelectric devices have the valuable property of being bistable. Once an applied electrical potential has altered the configuration,

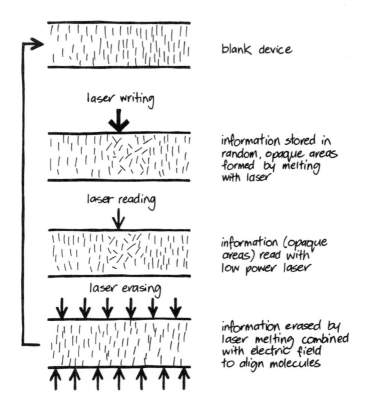

blank device

laser writing

information stored in random, opaque areas formed by melting with laser

laser reading

information (opaque areas) read with low power laser

laser erasing

information erased by laser melting combined with electric field to align molecules

Figure 15.3 Liquid crystal film as an information storage device.

one can turn off the signal without affecting the structure or the resulting model. This is so because cooperative rather than random movements of all the molecules are necessary for change[26].

In the production of oriented polymers such as the aromatic polyamide Kevlar® referred to above, sulphuric acid softens aromatic polyamides to yield a liquid crystal phase. This is spun into fibres before washing off the sulphuric acid. More recently, there has been greater interest in polymers that melt into a liquid crystal phase. This liquid crystal structure may be the result of the nature of either the main chain or backbone of the polymer molecules, or of the 'pendant' side-chain groups. Liquid crystal polymers provide reversible information storage (as in erasable compact discs) if domains can be both aligned electrically and randomised by laser melting (figure 15.3)[27].

Liquid crystal molecules show a tendency to align themselves because of their shape, but this occurs naturally (on a 'self-assembling materials' basis) only in microscopic domains. For practical applications like the twisted nematic liquid crystal displays (portable colour television and notebook computer screens) manufacturers have to achieve the macroscopic-scale uniform orientation by external means. The current technology (as much a craft as a science) confines the liquid crystal molecules between thin (one hundred nanometre) polymer films that are mechanically rubbed (with velvet fabric) to induce directionality. However, recent work shows that it is possible to macroscopically align liquid crystal molecules by means of polarised ultraviolet light[28].

15.9 Replication

The structure of 'life' depends on repeated molecular copying of information with a high level of precision. The DNA double-helix is the classic replicating structure, but there are simpler systems that exhibit aspects of replication. These involve chirality, 'autocatalysis' (where the products of a reaction act as catalysts, providing an acceleration of reaction rate) and polymerisation. The presence of one chiral defect in DNA (the replacement of one nucleotide by its enantiomer) introduces sufficient strain to prevent the formation of the hydrogen bonds in neighbouring groups. This disturbs its structure and functional ability, demonstrating the importance of chirality in biological replication[29]. Chirality is just one aspect of **order** in living matter[30].

James Lovelock[31] has emphasised in developing the Gaia model how processes of information transfer usually result in very rapid (exponential) attenuation over time and space. In contrast, molecular replication in living systems is a very efficient means of reliably conveying information. So although we do not usually think of the structure of materials in connection with information transfer, it does have this important role. The demands for new materials in information technology are driving many of the advances in the modelling of materials at the atomic level.

Associated with these ideas are the concepts of 'molecular machines', 'self-assembling materials', 'molecular architecture', 'antibody engineering', and the use of DNA to transmit chemical information and control the synthesis of materials from colloid systems. These are all aspects of our desire to design 'bottom-up' materials for particular purposes in which information content is just as important as structural properties[32].

15.10 Biological Matter

Most carbon compounds are described as 'organic'. This term dates from the time when chemical compounds originating in living systems were thought to have some unique quality not present in other chemical compounds. Chemists soon abandoned this model because they found they could synthesise many of these 'natural' compounds and rearrange even the most complex.

The most important biological material is the apparently simple compound water. Although we usually think of water as 'inorganic' rather than 'organic', when water acts as a solvent it is capable of a wide variety of specific and subtle interactions in biological systems.

The efficiencies of chemical processes in biological systems still surprise chemists. Recent work[33] has shown that in the oxygen-transporting protein haemoglobin the central region of this complex structure opens and closes like a clam shell. This can trap and release an oxygen molecule within a few thousandths of a nanosecond.

The inanimate materials around us can serve as models of the often more complex materials resulting from biological processes. Liquid crystals and surfactants have much in common with living cells[34], and the techniques of x-ray crystallography are just as applicable to protein crystals as to inorganic crystals. Computer-generated images have greatly helped chemists and biochemists trying to visualise structures and their transformations[35]. Enzymes are the ultimate in catalysts but the mathematical models and mechanistic models used for enzymes in chemical reaction kinetics are exactly the same as those for 'simpler' systems. Nature makes great use of chirality, but chemists are starting to learn these tricks as well[36]. Although we have become proficient at synthesising and modifying the chemical **compounds** of biological systems, we cannot say the same of natural biological **structures** or 'biocomposites'. These are hierarchical composite structures with specialised functional achievements, such as[37]:

- the lignocellulose of plants,
- the collagen of animal tendons,
- the combination of collagen and hydroxyapatite (calcium carbonate) in bone.

It is common practice to model these structures as organised on at least five distinct, successive levels:

- atomic,
- molecular or nanoscopic,
- microscopic,
- mesoscopic,
- macroscopic.

These 'artificial' distinctions have much in common with the way we classify other fractal systems that are actually continuous, like the blood distribution system in the body, into discrete structural parts. However, in some situations it may prove instructive to think in terms of a continuous structure from the atomic to the macroscopic scale. Intermolecular covalent bonds exist between specific sites, and nanoscale van der Waals interactions link the various levels. Our attempts at synthetic composites (such as steel-reinforced concrete, plywood and fibre bundles) are primitive by comparison with these biocomposites.

The distinction between 'organic' and 'inorganic' materials is becoming less clear, as illustrated in the Gaia model[38]. As an example, the element calcium is toxic within living cells at levels above a few parts per million despite being an essential component of many physiological systems. The ability of organisms to precipitate calcium carbonate both increased the chance of survival of these organisms and caused the deposition of large amounts of calcium as calcium carbonate.

I have referred to the philosophy of self-assembling materials in several contexts, such as the fullerene nanoparticles. One can describe the particular self-assembly control with one of several models depending on the particular process:

- thermodynamic, for equilibrium control,

- kinetic, for time-dependent control, or

- fractal, for chaotic control.

In all cases it is useful to have ready-made biological examples when attempting synthetic processes[39].

15.11 Biomolecular Machines

A living organism is an assembly of molecular-scale machines that converts energy from chemicals (or light) into movement, heat and internal construction and repair (figure 15.4). We are now able to build synthetic biomolecular machines that model the behaviour of natural ones.

One example is a rubber-like elastic protein polymer material that mimics elastin, the elastic protein occurring in various parts of our bodies. Studies of this kind not only provide information on how natural mechanical biological systems work, but also have

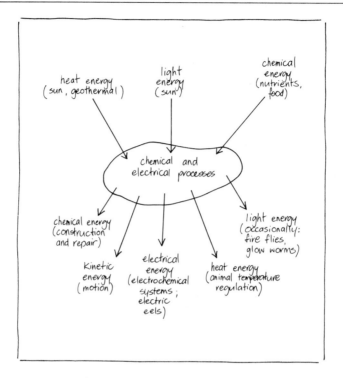

Figure 15.4 A living organism is a biomolecular machine converting energy from one form to another.

extensive applications, both medical and nonmedical. For example, consider a biomolecular desalination material that contracts in the presence of salt. A mechanical device could stretch it in salt water to absorb pure water, then squeeze out the pure water. We can discuss the behaviour of biomolecular machines that convert heat into work with the models I have introduced in previous chapters:

- First use the molecular model of the protein polymer to show how the stretching of the polymer chain causes twisting of the side chains and increased ordering of the molecular array.

- Then invoke the entropy model in discussing how the natural drive towards disorder causes the polymer to contract back to its original configuration after we release it.

- Classify different parts of the protein polymer as hydrophobic or hydrophilic (terms introduced above), so their interactions with water change depending on the stretch status of the protein.

- The hydrogen-bonding model accounts for the observation that the hydrophobic parts of the protein and the water molecules interact to form an extended array. At higher temperatures this breaks down, allowing the protein to become folded.

The results from this combination of models are consistent with the observation that the polymers contract and so convert heat into work. It is in this process of converting heat into work where engineers have been fairly clever at using mechanical machines (although, of course, living organisms were doing it first)[40].

15.12 Life

In terms of the thermodynamic model, the process described as 'life' is a most improbable event, a local system characterised by marked entropy decrease supported by the entropy increase of its surroundings. It is the most important of the processes described as 'dissipative'[41]; life is a movement 'upstream' against the flow of thermodynamically 'natural' processes that are moving inevitably towards equilibrium and 'death'. Living things excrete waste products (both chemical material and infrared energy) high in entropy. Organisms, which are very complex, can be thought of as a concentrated source of information, a view that is also consistent with the alternative definition of entropy occurring in information science[42].

James Lovelock developed a new perspective on the evolutionary process on Earth that encompassed and extended the adaptation model of Charles Darwin. He considered life to be a planetary phenomenon where not only do better adapted organisms pass on these qualities to offspring, but also where the growth of the organisms affects the physical and chemical environment. This whole-Earth model he called Gaia (the Greek name for the Earth Goddess). Within the scientific discipline 'geophysiology', he integrated physical, chemical, biological and geological systems on Earth into a single, indivisible process. Such a model is easier to accept when you realise that a tree, which one tends to think of as a living entity, is a thin layer of living cells making up perhaps no more than a few per cent of its mass supported by the dead structure of its heartwood. While developing the Gaia model, Lovelock made use of simpler whole-Earth models such as Daisyworld (a world inhabited only by two forms of daisy) that were amenable to quantitative mathematical investigation[43].

This is not the only model for the way the Earth's life-supporting environment and climate developed. Other scientists consider that the fundamental feedback controls on levels of carbon dioxide in the carbon cycle and other critical features are physical and chemical rather than biological. It follows that the Earth would have developed habitable conditions even in the absence of inhabitants. We have the choice of accepting one or other of these models, or of adopting some intermediate position[44].

15.13 Evolving Universes

Rupert Sheldrake[45] developed the holistic concept of a collective Universal 'superorganism' in which both physical processes and human thought processes participate. Also as a result of this structure, information is transferred and ideas evolve. While this view ('morphic resonance') is not widely held, we should remain receptive to any new model that is consistent with known facts. But the most ambitious of our array of models is due to physicists like Lee Smolin (Pennsylvania State University), Andrei Linde (Lebedev Physics Institute, Moscow) and Stephen Hawking (Cambridge University) who are moving towards the concept of 'evolving universes'.

In this model, black holes generate new embryonic, growing universes. Each emerging universe has its own particular values for the 'fundamental' physical properties (like the gravitational constant). Because these values are slightly different, there is the opportunity for universes to evolve by a natural selection mechanism. The more 'appropriate' or 'successful' the values of these constants, the longer that particular universe survives and the greater the chance it has of budding off new universes from a larger number of black holes. This parallels the biological evolutionary model, resulting in the most 'successful' universes surviving at the expense of less successful ones. Presumably the successful universes must have physical characteristics similar to those of our own universe. This model 'explains' why the expansion rate of our universe has been 'just right' to allow stars and galaxies to form[46].

Endnotes

[1]Further reading Witten (1990), de Gennes (1992), Lubensky and Pincus (1984). See also **1.8 Structure**.

[2]See **6.5 Failure**.

[3]See **11.5 Glasses**.

[4]Further reading de Gennes (1984).

[5]See **3.5 Flow**, **11.4 Viscous Flow** and **15.4 Flowing Polymers**.

[6]Further reading Volkmuth and Austin (1992).

[7]See also **13.5 Organic Conductors**. Further reading Flam (1991).

[8]See **5.2 Thermodynamic Model**.

[9]See **12.4 Phase Diagrams** and **12.6 Cohesion Parameters**. Further reading De Simone *et al* (1992), Barton (1983a, 1985, 1991).

[10]Further reading Bird and Curtiss (1984), Evans *et al* (1984).

[11]Further reading Bird and Curtiss (1984).

[12]Further reading Witten (1990).

[13]Further reading Whittle and Bullough (1992), Halsey and Martin (1993).

[14]Further reading Petford (1991, p 41).

[15]Sometimes it is possible to remove the liquid without the polymer structure collapsing, to form aerogels (**14.10 Aerogels**).

[16]See **11.4 Viscous Flow**.

[17]Further reading Osada and Ross-Murphy (1993), Tanaka (1981). See also **2.4 Condition Critical** and **15.11 Biomolecular Machines**.

[18]See **1.14 How Small Are Atoms?** and **2.1 Scientific Jargon**.

[19]See **14.13 Aggregates** and **3.8 Fractals**. Further reading Witten (1990).

[20]Further reading Lubensky and Pincus (1984), Witten (1990). See also **13.9 Magnetic Matter** and **14.7 Surfactants**.

[21]Further reading Seddon and Templer (1991), Collings (1990). See also **6.2 Probing Matter**.

[22]Further reading Coghlan (1991), *Education in Chemistry* (1991), Collings (1990). See also **9.12 Handedness**.

[23]In connection with superfluids: see **11.3 Superfluid Helium**.

[24]Further reading Brock *et al* (1989), Templer and Attard (1991), de Gennes (1992).

[25]Further reading Seddon and Templer (1991), de Gennes (1992). See also **14.7 Surfactants**.

[26]Further reading Templer and Attard (1991).

[27]Further reading Attard and Imrie (1991). See also **15.2 Polymers**.

[28]Further reading van Haaren (1996).

[29]Further reading Orgel (1992), Avetisov *et al* (1991).

[30]Kauffman (1993).

[31]Lovelock (1988, p 67).

[32]*Chemistry in Australia* (1992), Bethell and Schiffrin (1996).

[33]Hecht (1991a).

[34]Further reading Seddon and Templer (1991).

[35]For x-ray crystallography see Chapter 6. Further reading McPherson (1989), Olson and Goodsell (1992).

[36]See **12.8 Chemical Processes** and **9.12 Handedness**.

[37]Further reading Gordon (1968), Baer *et al* (1992).

[38]Further reading Lovelock (1988).

[39]See **9.14 Fullerenes**, **5.2 Thermodynamic Model**, **12.8 Chemical Processes** and **3.8 Fractals**. Further reading Whitesides (1995).

[40]Further reading Urry (1996).

[41]See **5.7 Ordering**. Further reading Davies (1980, Chapter 8).

[42]See **5.4 Energy and Entropy**.

[43]Further reading Lovelock (1988, 1991).

[44]Further reading Kasting *et al* (1988).

[45]Sheldrake (1987).

[46]Further reading Gribbin (1994).

POSTSCRIPT

STATES OF MIND

If you return to the question I asked in Chapter 1 concerning the number of states of matter, it should now be apparent that the number of possible states of matter is limited only by our ability to perceive them. They include not only the materials available to us now, and those waiting to be found, but also those not yet in existence in the Universe. Your imagination, your state of mind, imposes the only limits.

BIBLIOGRAPHY

Adams G B, Sankey O F, Page J B, O'Keeffe M and Drabold D A 1992 Energetics of Large Fullerenes: Balls Tubes and Capsules *Science* **256** 1792–5

Adams S F 1992 Beta Rays and Neutrons *Physics Education* **27** 102–8

Aihara J-I 1992 Why Aromatic Compounds are Stable *Scientific American* **226** (3) 44–50

Ajayan P M and Iijima S 1992 Smallest Carbon Nanotube *Nature* **358** 23

Alder B J and Alley W E 1984 Generalized Hydrodynamics *Physics Today* **37** (1) 56–63

Aldous P 1992 Making Buckyballs Go Ballistic *Science* **257** 1481

Amaratunga G A J, Chhowalla M, Kiely C J, Alexandrou I, Aharanov R and Devenish R M 1996 Hard Elastic Carbon Thin Films from Linking of Carbon Nanoparticles *Nature* **383** 321–3

Amato I 1992a A First Sighting of Buckyballs in the Wild *Science* **257** 167

——1992b A New Blueprint for Water's Architecture *Science* **256** 1764

——1992c Looking Glass Chemistry *Science* **256** 964–6

——1992d Microgravity Materials Science Strives to Stay in Orbit *Science* **257** 882–3

——1992e The Ascent of Odorless Chemistry *Science* **256** 306–8

Apfel R E 1972 The Tensile Strength of Liquids *Scientific American* **227** (6) 58–71

Arnau A, Tunon I and Silla E 1995 The Discovery of the Chemistry Among the Stars *Journal of Chemical Education* **72** (9) 776–81

Ashkin A 1972 The Pressure of Laser Light *Scientific American* **226** (2) 63–71

Atkins P W 1984 *The Second Law* (New York: Scientific American Books, Freeman) (1994 *The Second Law: Energy, Chaos and Form* Revised edn (New York: Scientific American Library, Freeman))

——1990 *Physical Chemistry* 4th edn (Oxford: Oxford University Press)

Attard G and Imrie C 1991 Plastics of a New Order *New Scientist* **130** (1768, 11 May) 34

Aubert J H, Kraynik A M and Rand P B 1986 Aqueous Foams *Scientific American* **254** (5) 58–66

Auel J M 1991 *The Plains of Passage* (London: Coronet)

Austin S M and Bertsch G F 1995 Halo Nuclei *Scientific American* **272** (6) 62–7

Avetisov V A, Goldanksii V I and Kuz'min V K 1991 Handedness Origin of Life and Evolution *Physics Today* **44** (7) 33–41

Baer E, Hiltner A and Morgan R J 1992 Biological and Synthetic Hierarchical Composites *Physics Today* **45** (10) 60–7

Baggott J 1990 Quantum Mechanics and the Nature of Physical Reality *Journal of Chemical Education* **67** 638–42

——1992 Beating the Uncertainty Principle *New Scientist* **130** (1808, 15 February) 26–30

Bak P and Paczuski M 1993 Why is Nature so Complex? *Physics World* **6** (12) 39–43

Ball P 1996 The Perfect Nanotube (Review of *Fullerenes '96* Conference, Oxford, July 1996) *Nature* **382** 207–8

Ball P and Garwin L 1992 Science at the Atomic Scale *Nature* **355** 761–6

Bardeen J 1990 Superconductivity and Other Macroscopic Quantum Phenomena *Physics Today* **43** (12) 25–31

Barlow R 1992 Particle Physics: from School to University *Physics Education* **27** 92–5

Barnes G T 1992 New Techniques for the Study of Monolayers at the Air/Water Interface *Chemistry in Australia* **59** (10) 532

Barta N S and Stille J R 1994 Grasping the Concepts of Stereochemistry *Journal of Chemical Education* **71** 20–23

Barton A F M 1973 The Description of Dynamic Liquid Structure *Journal of Chemical Education* **50** 119

——1974 *The Dynamic Liquid State* (London: Longman)

——1975 Solubility Parameters *Chemical Reviews* **75** 731

——1982 The Application of Cohesion Parameters to Wetting and Adhesion *Journal of Adhesion* **14** 33

——1983a Cohesion Parameters *Polymer Yearbook* ed H G Elias and R A Pethrick (Chur: Harwood) pp 149–82

——1983b *Handbook of Solubility Parameters and Other Cohesion Parameters* (Boca Raton, FL: CRC Press)

——1985 Applications of Solubility Parameters and Other Cohesion Parameters in Polymer Science and Technology *Pure and Applied Chemistry* **57** 905–12

——1987 Cohesion Parameters *Encyclopaedia of Physical Science and Technology* vol 3 (New York: Academic) pp 140–56

——1990 *Handbook of Polymer–Liquid Interaction Parameters and Solubility Parameters* (Boca Raton, FL: CRC Press)

——1991 *Handbook of Solubility Parameters and Other Cohesion Parameters* 2nd edn (Boca Raton, FL: CRC Press)

Barton A F M and Hodder A P W 1973 Explosive Phase Transitions

Chemical Reviews **73** 127

Barton A F M, Hodder A P W and Wilson A T 1971 Explosive or Detonative Phase Transitions on a Geological Scale *Nature* **234** 293

Bate R T 1988 The Quantum Effect Device: Tomorrow's Transistor? *Scientific American* **258** (3) 78–82

Batlogg B 1991 Physical Properties of High-T_c Superconductors *Physics Today* **44** (6) 44–50

Beaton J M 1992 A Paper-Pattern System for the Construction of Fullerene Molecular Models *Journal of Chemical Education* **69** 610–2

Beaton J M 1995 Paper Models for Fullerenes from C_{60}–C_{84} *Journal of Chemical Education* **72** 863–9

Belak J F 1993a Nanotribology *MRS Bulletin* **18** (5) 15–7

——1993b Simulation of Nanometer-Scale Deformation of Metallic and Ceramic Surfaces *MRS Bulletin* **18** (5) 55–60

Benkovic S J 1996 The Key is in the Pocket *Nature* **383** (6595, 5 September) 23–4

Bennett C H 1992 Quantum Cryptography: Uncertainty in the Service of Privacy *Science* **257** 752–3

Bennett C H, Brassard G and Eckert A K 1992 Quantum Cryptography *Scientific American* **267** (4) 26–33

Bernstein J 1996 The Reluctant Father of Black Holes *Scientific American* **274** (6) 66–72

Berry M 1990 Anticipations of the Geometric Phase *Physics Today* **43** (12) 34–40

Berry M V, Percival K and Weiss N O (ed) 1987 Dynamical Chaos *Proceedings of the Royal Society London* A **413** (1844, 8 September) 1–199

Berry R S 1990 When the Melting and Freezing Points Are Not the Same *Scientific American* **263** (2) 50–6

Bethell D and Schiffrin D J 1996 Nanotechnology and Nucleotides *Nature* **382** 581

Binnig G and Rohrer H 1985 The Scanning Tunneling Microscope *Scientific American* **253** (2) 40–6

Bird R B and Curtiss C F 1984 Fascinating Polymeric Liquids *Physics Today* **37** (1) 36–43

Bishop D J 1993 Heroic Holograms *Nature* **366** 209

Bishop D J, Gammel P L and Huse D A 1993 Resistance in High-Temperature Superconductors *Scientific American* **268** (2) 24–31

Bollinger J J and Wineland D J 1990 Microplasmas *Scientific American* **262** (1) 114–20

Boo W O J 1992 An Introduction to Fullerene Structures. Geometry and Symmetry *Journal of Chemical Education* **69** 605–9

Borra E F 1994 Liquid Mirrors *Scientific American* **270** (2) 50–5

Bracewell R N 1989 The Fourier Transform *Scientific American* **260** (6) 62–9

Braun R D 1992 Scanning Tunneling Microscopy of Silicon and Carbon *Journal of Chemical Education* **69** A90–2

Brock J D, Birgeneau R J, Lister J D and Aharoni A 1989 Liquids Crystals and Liquid Crystals *Physics Today* **42** (7) 52–9

Brumer P and Shapiro M 1995 Laser Control in Chemical Reactions *Scientific American* **272** (3) 34–9

Buck B, Merchant A C and Perez S M 1993 An Illustration of Benford's First Digit Law Using Alpha Decay Half Lives *European Journal of Physics* **14** 59–63

Buseck P R, Tsipursky S J and Hettich R 1992 Fullerenes from the Geological Environment *Science* **257** 215–7

Cahn R W 1996 The Baking of Layer Cakes *Nature* **382** 405–6

Canham L 1993 A Glowing Future for Silicon *New Scientist* **138** (1868, 10 April) 23–7

Capasso F, Sirtori C, Faist J, Sivco D L, Chu S-N G and Cho A Y 1992 Observation of an Electronic Bound State above a Potential Well *Nature* **358** 565–7

Cardy J 1993 Conformal Field Theory Comes of Age *Physics World* **6** (6) 29–33

Cava R J 1990 Superconductors Beyond 1-2-3 *Scientific American* **263** (2) 24–31

Chang L L and Esaki L 1992 Semiconductor Quantum Heterostructures *Physics Today* **45** (10) 36–43

Chaudhury M K and Whitesides G M 1992 How to Make Water Run Uphill *Science* **256** 1539–41

Chemistry in Australia 1992 Molecular Engineering Feature **59** 572–88

Chen C T, Tjeng L H, Rudolf P, Meigs G, Rowe J E, Chen J, McCauley J P Jr, Smith A B III, McGhie A R, Romanow W J and Plummer E W 1991 Electronic States and Phases of K_xC_{60} from Photoemission and X-ray Absorption Spectroscopy *Nature* **352** 603–5

Chiao R Y, Kwiat P G and Steinberg A M 1993 Faster than Light? *Scientific American* **269** (2) 38–46

Chu S 1992 Laser Trapping of Neutral Particles *Scientific American* **266** (2) 49–54

Cline D B 1994 Low-Energy Ways to Observe High-Energy Phenomena *Scientific American* **271** (3) 26–33

Coghlan A 1991 Clothes that Change Colour in the Heat of the Moment *New Scientist* **130** (1768, 11 May) 21

Cohen E G D 1984 The Kinetic Theory of Fluids—an Introduction *Physics Today* **37** (1) 64–73

Collings P J 1990 *Liquid Crystals. Nature's Delicate Phase of Matter*

(Bristol: Adam Hilger)

Collins G P 1992 Quantum Cryptography Defies Eavesdropping *Physics Today* **45** (11) 21–3

Cook N and Dallacasa V 1988 A Crystal Clear View of the Nucleus *New Scientist* **117** (1606, 31 March) 44–6

Corcoran E 1990 Trends in Materials: Diminishing Dimensions *Scientific American* **263** (5) 74–83

Craig D P 1992 Light Crystal Forces and Photochemistry *Chemistry in Australia* **59** 651–3

Craig N C, Gee G C and Johnson A R 1992 C_{60} and C_{70} Made Simple *Journal of Chemical Education* **69** 664–6

Crawford H J and Greiner C H 1994 The Search for Strange Matter *Scientific American* **270** (1) 58–63

Crutchfield J P, Farmer J D, Packard N H and Shaw R S 1986 Chaos *Scientific American* **255** (6) 38–49

Curl R F 1992 C_{60} Yesterday and Today *ChemNZ* (47) 3–14

Curl R F and Smalley R E 1991 Fullerenes *Scientific American* **265** (4) 32–40

Dasent W E 1963 Non-existent Compounds *Journal of Chemical Education* **40** 130

——1965 *Nonexistent Compounds: Compounds of Low Stability* (New York: Marcel Dekker)

Davies P 1982 *Other Worlds* (London: Sphere)

—— *Superforce* (London: Unwin)

de Gennes P G 1984 Entangled Polymers *Physics Today* **36** (6) 33

——1992 Soft Matter *Science* **256** (5056, 24 April) 495–7

De Simone J M, Zihibin Guan and Elsbernd C S 1992 Synthesis of Fluoropolymers in Supercritical Carbon Dioxide *Science* **257** 945–7

Dewdney A K 1987 Computer Recreations: Beauty and Profundity: the Mandelbrot Set and a Flock of its Cousins called Julia *Scientific American* **257** (5) 118–22

——1989 Computer Recreations: a Tour of the Mandelbrot Set Aboard the Mandelbus *Scientific American* **260** (2) 88–91

Ditto W L and Pecora L M 1993 Mastering Chaos *Scientific American* **269** (2) 62–8

Dobson K 1993 The Education Time Tunnel *Physics World* **6** (5) 15–6

Donn B D 1990 The Formation and Structure of Fluffy Cometary Nuclei from Random Accumulation of Grains *Astronomy and Astrophysics* **235** 441–6

Donnelly R J 1988 Superfluid Turbulence *Scientific American* **259** (5) 66–74

——1995 The Discovery of Superfluidity *Physics Today* **48** (7) 30–6

Dresselhaus M S 1992 Down the Straight and Narrow *Nature* **358** 195–6

Duncan M A and Rouvray D H 1989 Microclusters *Scientific American* **261** (6) 60–5

Earnshaw R 1993 Scientific Visualization: the State of the Art *Physics World* **6** (9) 48–51

Eaves L 1992 Looking inside Quantum Dots *Nature* **357** 540

Ebbesen T W and Ajayan P M 1992 Large-Scale Synthesis of Carbon Nanotubes *Nature* **358** 220–2

Ebenezer J V 1992 Making Chemistry Learning More Meaningful *Journal of Chemical Education* **69** 464–7

Education in Chemistry 1991 Chemically Chic Liquid Crystals **28** (4) 91

Eigler D M, Lutz C P and Rudge W E 1991 An Atomic Switch Realized with the Scanning Tunnelling Microscope *Nature* **352** 600–3

Ekstrom P and Wineland D 1980 The Isolated Electron *Scientific American* **243** (2) 90

Elliott S R 1991 Medium-Range Structural Order in Covalent Amorphous Solids *Nature* **354** 445–52

Emsley J 1993 The Weird and Wonderful World of Buckyballs *New Scientist* **138** (1872, 8 May) 13

Englert B-G, Scully M O and Walther H 1994 The Duality in Matter and Light *Scientific American* **271** (6) 86–92

Evans D J, Hanley H J M and Hess S 1984 Non-Newtonian Phenomena in Simple Fluids *Physics Today* **37** (1) 26–33

Fagan P J and Ward M D 1992 Building Molecular Crystals *Scientific American* **267** (1) 28

Falicov L M 1992 Metallic Magnetic Superlattices *Physics Today* **45** (10) 46–51

Farmelo G 1992 Teaching Particle Physics in The Open University's Science Foundation Course *Physics Education* **27** 96–101

Feynman R P 1985 *QED: The Strange Theory of Light and Matter* (Princeton, NJ: Princeton University Press)

Fineberg J 1996 Physics in a Jumping Sandbox *Nature* **382** 763–4

Finnis M 1993 Simulation on All Length Scales *Physics World* **6** (7) 37–42

Flam F 1991 Plastics Get Oriented—and Get New Properties *Science* **251** 874–6

——1992 Physicists Rock the Standard Model in Dallas *Science* **257** 1044–5

Fortman J J 1993 Pictorial Analogies I: States of Matter; II: Types of Solid; IX: Liquids and Their Properties *Journal of Chemical Education* **70** 56–7, 57–8, 881–2

——1994 Pictorial Analogies X: Solutions of Electrolytes *Journal of Chemical Education* **71** 27–8

Fricke J 1988 Aerogels *Scientific American* **258** (5) 68

Friend C M 1993 Catalysis on Surfaces *Scientific American* **268** (4) 42–7

Ge M and Sattler K 1993 Vapor-Condensation Generation and STM Analysis of Fullerene Tubes *Science* **260** 515–8

Geake E 1994 Colourful Glass Puts Pollutants to the Test *New Scientist* **141** (1907, 8 January) 15

Geis M W and Angus J C 1992 Diamond Film Semiconductors *Scientific American* **267** (4) 64–9

Geselbracht M J, Ellis A B, Penn R L, Lisensky G C and Stone D S 1994 Mechanical Properties of Metals *Journal of Chemical Education* **71** 254–61

Gillan M 1993 Gulliver among the Atoms *New Scientist* **138** (1867, 3 April) 34–7

Gillespie R J, Spencer J N and Moog R S 1996 Demystifying Introductory Chemistry. Part 2. Bonding and Molecular Geometry Without Orbitals—The Electron Domain Model *Journal of Chemical Education* **73** 622–7

Girvin S M 1992 Anyons Superconduct But Do Superconductors Have Anyons? *Science* **257** 1354–5

Gleick J 1988 *Chaos: Making a New Science* (London: Sphere)

Goldman T, Hughes R J and Nieto M M 1988 Gravity and Antimatter *Scientific American* **258** (3) 32–40

Goldstein A N, Echer C M and Alivisatos A P 1992 Melting in Semiconductor Nanocrystals *Science* **256** 1425–7

Goodman W D 1990 On Knowledge and Memory *Journal of Chemical Education* **67** 678–9

Goodstein D L 1975 *States of Matter* (Englewood Cliffs, NJ: Prentice-Hall)

Gordon J E 1968 *The New Science of Strong Materials* (Harmondsworth: Penguin) and also 2nd edn

Gorin G 1994 Mole and Chemical Amount *Journal of Chemical Education* **71** 114–6

——1996 Mendeleev and Moseley. The Principal Discoverers of the Periodic Law *Journal of Chemical Education* **73** 490–3

Graham G R 1991 Let's See It for Real: a New Medium for an Old Message *Physics Education* **26** 355–8

Greaves R J and Schlecht K D 1992 Gibbs Free Energy: the Criteria for Spontaneity *Journal of Chemical Education* **69** 417–8

Gribbin J 1994 Is the Universe Alive? *New Scientist* **141** (1908, 15 January) 38–40

Guo Y, Karasawa N and Goddard W A III 1991 Prediction of Fullerene Packing in C_{60} and C_{70} Crystals *Nature* **351** 464

Gutzwiller M C 1990 *Chaos in Classical and Quantum Mechanics* (New York: Springer)
——1992 Quantum Chaos *Scientific American* **262** (1) 26–32

Halsey T C and Martin J E 1993 Electrorheological Fluids *Scientific American* **269** (4) 42–8
Hammond G S 1992 The Fullerenes: Overview 1991 *Fullerenes: Synthesis Properties and Chemistry of Large Carbon Clusters* ed G S Hammond and V J Kuck (*ACS Symposium Series* **481**) pp ix–xiii
Hammond G S and Kuck V J (ed) 1992 *Fullerenes: Synthesis Properties and Chemistry of Large Carbon Clusters* (*ACS Symposium Series* **481**)
Hanley H J M 1984 Fluids out of Equilibrium. Introduction to series of articles by Evans *et al*, Bird and Curtiss, Alder and Alley, and Cohen in special issue *Physics Today* **37** (1) 25
Haroche S and Raimond J-M 1993 Cavity Quantum Electrodynamics *Scientific American* **268** (4) 26–33
Harrison A M 1993 The Interpretation of Quantum Theory to Beginners *Journal of Chemical Education* **70** 260
Harrison J A, White C T, Colton R J and Brenner D W 1993 Atomistic Simulations of Friction at Sliding Diamond Interfaces *MRS Bulletin* **19** (5) 50–3
Hawking S W 1988 *A Brief History of Time* (Ealing: Bantam, Transworld)
Hawking S W and Penrose R 1996 The Nature of Space and Time (extract from a book of the same name published by Princeton University Press, 1996, based on a 1994 lecture series at the Isaac Newton Institute for Mathematical Sciences at Cambridge University) *Scientific American* **275** (1) 44–9
Hayward A T J 1971 Negative Pressure in Liquids: Can It be Harnessed to Serve Man? *American Scientist* **59** 434–43
Hazen R M 1988 Perovskites *Scientific American* **258** (6) 52–61
Healy T W 1992 Colloid Science and Mineral Processing *Chemistry in Australia* **59** 510–2
Hebard A F 1992 Superconductivity in Doped Fullerenes *Physics Today* **45** (11) 26–32 and front cover
Hecht J 1991a Interfering Atoms Feel a Sense of Acceleration *New Scientist* **130** (1774, 22 June) 18
——1991b Soccer-Ball-Shaped Molecule Becomes a Superconductor *New Scientist* **130** (1766, 27 April) 15
Hefter G T, Barton A F M and Chand A 1991 Semi-Automated Apparatus for the Determination of Liquid Solubilities: Mutual Solubilities of Water and Butan-2-ol *Journal of the Chemical Society, Faraday Transactions* **87** 591–7

Hegstrom R A and Kondepudi D K 1990 The Handedness of the Universe *Scientific American* **262** (1) 98–105

Heilbronner E and Dunitz J D 1993 *Reflections on Symmetry in Chemistry and Elsewhere* (Weinheim: VCH)

Heller E J 1996 Quantum Chaos for Real *Nature* **380** 583–4

Heller E J, Crommie M F, Lutz C P and Eigler D M 1994 Scattering and Absorption of Surface Electron Waves in Quantum Corrals *Nature* **369** 464–6

Herman H 1988 Plasma-Sprayed Coatings *Scientific American* **259** (3) 78–83

Hoffmann R 1993 How Should Chemists Think? *Scientific American* **268** (2) 40–9

Holderness M 1994 Is Science Habit-Forming? (Review of *Paradigms and Barriers: How Habits of Mind Govern Scientific Beliefs* by Howard Margolis (Chicago, Il: University of Chicago Press)) *New Scientist* **141** (1907, 8 January) 42

Hollfelder F, Kirby A J and Tawfik D S 1996 Off-the-Shelf Proteins that Rival Tailor-Made Antibodies as Catalysts *Nature* **383** 60–3

Home D and Gribbin J 1994 The Man Who Chopped Up Light *New Scientist* **141** (1907, 8 January) 26–9

Hoover W G 1984 Computer Simulation of Many-Body Dynamics *Physics Today* **37** (1) 44–50

Horgan J 1992 Quantum Philosophy *Scientific American* **267** (1) 72–80

——1995 Particle Metaphysics *Scientific American* **270** (2) 70–8

——1996 *The End of Science* (New York: Addison-Wesley Longman)

Huffman D R 1991 Solid C_{60} *Physics Today* **44** (11) 22–9

Hull D L 1996 A Revolutionary Philosopher of Science *Nature* **382** 203–4

Huse D A, Fisher M P A and Fisher D S 1992 Are Superconductors Really Superconducting? *Nature* **358** 553–9

Iacoe D W, Potter W T and Teeters D 1992 Simple Generation of C_{60} (Buckminsterfullerene) *Journal of Chemical Education* **69** 663

Jaeger H M and Nagel S R 1992 Physics of the Granular State *Science* **255** 1523–31

Jennings B R and Morris V J 1974 *Atoms in Contact* (Oxford Physics Series)(Oxford: Clarendon)

Jeon D, Kim J, Gallagher M C and Willis R F 1992 Scanning Tunneling Spectroscopic Evidence for Granular Metallic Conductivity in Conducting Polymeric Polyaniline *Science* **256** 1662–4

Jewell J L, Harbison J P and Scherer A 1991 Microlasers *Scientific American* **265** (5) 56–62

Jones D G C 1991 Teaching Modern Physics—Misconceptions of the Photon that can Damage Understanding *Physics Education* **26** 93–8

Jones G T 1992 A Quite Extraordinary Demonstration of Quantum Electrodynamics *Physics Education* **27** 81–6

Jones T and McConville C 1995 Atomic Views of Semi-Conductor Surfaces *Physics World* **8** (1) 35–40

Jorgensen J D 1991 Defects and Superconductivity in the Copper Oxides *Physics Today* **44** (6) 34–40

Journal of Chemical Education 1992 The Fullerenes: a Synthesis of Chemistry and Aesthetics **69** 604

Journal of Chemical Education 1994 In this Issue: Organizing Information for the Human Mind **71** 2

Kadanoff L P 1986 Fractals: Where's the Physics? *Physics Today* **39** (2) 6–7

——1991 Complex Structures from Simple Systems *Physics Today* **44** (3) 9–10

Kaner R B and MacDiarmid A G 1988 Plastics that Conduct Electricity *Scientific American* **258** (2) 60–5

Kasting J F, Toon O B and Pollack J B 1988 How Climate Evolved on the Terrestrial Planets *Scientific American* **258** (2) 46–53

Kastner M A 1993 Artificial Atoms *Physics Today* **46** (1) 24–31

Kauffman S A 1993 *The Origins of Order: Self-Organization and Selection in Evolution* (New York: Oxford University Press)

Kaye B 1993 *Chaos and Complexity: Discovering the Surprising Patterns of Science and Technology* (Weinheim: VCH)

Kelly P M (ed) 1981 *Materials for the Future* (Papers Meeting of Science and Industry Forum Australian Academy of Sciences March 1979) Forum Report No 14 1980 (Canberra: Australian Academy of Sciences) particularly A K Head 'Why Materials Behave as they Do' pp 1–8

Kendall K 1980 Sticky Solids *Contemporary Physics* **21** 277–97

Kerr R A 1992 From Mercury to Pluto Chaos Pervades the Solar System *Science* **257** 33

Kivelson S, Lee D-H and Zhang S-C 1996 Electrons in Flatland *Scientific American* **274** (3) 64–9

Klemperer W 1992 Intermolecular Interactions *Science* **257** 887–8

Kleppner D, Littman M G and Zimmerman M L 1981 Highly Excited Atoms *Scientific American* **244** (5) 108–22

Kolb D 1977 But if Atoms are So Tiny *Journal of Chemical Education* **54** 543–7

Kuhn T S 1970 *The Structure of Scientific Revolutions* 2nd edn (Chicago, IL: University of Chicago Press)

Lambourne R 1992 Predicting the Physics of Particles *Physics Education* **27** 71–5

Landman U and Luedtke W D 1993 Interfacial Junctions and Cavitation *MRS Bulletin* **19** (5) 36–44

Langreth R 1993 Molecules Building Blocks of 21st Century Machines *The Australian* 27 April, pp 30–1

Leatherdale W H 1974 *The Role of Analogy Model and Metaphor in Science* (Amsterdam: North-Holland)

Likharev K K and Claeson T 1992 Single Electronics *Scientific American* **266** (6) 50–5

Lisensky G C, Kelly T F, Neu D R and Ellis A B 1991 The Optical Transform *Journal of Chemical Education* **68** 91–6

Lounasmaa O V and Pickett G 1990 The ^{3}He Superfluids *Scientific American* **262** (6) 64–71

Lovelock J 1988 *The Ages of Gaia* (Oxford: Oxford University Press)

——1991 *Gaia: The Practical Science of Planetary Medicine* (North Sydney: Allen and Unwin North)

Lubensky T C and Pincus P A 1984 Superpolymers Ultraweak Solids and Aggregates *Physics Today* **37** (10) 44–50

Maddox J 1992a Is Charge Quantization Exact? *Nature* **358** 449

——1992b Why Pebbles Float to the Surface *Nature* **358** 535

——1993 Calculating the Energy of Fullerenes *Nature* **363** 395

Main P 1993 When Electrons Go with the Flow *New Scientist* **138** (1877, 12 June) 30–3

Mandelbrot B B 1983 *The Fractal Geometry of Nature* (New York: Freeman)

McBride J M 1992 Crystal Polarity: a Window on Ice Nucleation *Science* **256** 814

McPherson A 1989 Macromolecular Crystals *Scientific American* **260** (3) 42–9

Mehta A and Barker G 1991 The Self-Organising Sand Pile *New Scientist* **130** (1773, 15 June) 34

Moffat A S 1992 New Methods Make Mid-Sized Molecules Easier to Solve *Science* **256** 309–10

Morgan N 1993 Electron Avalanche Aids Photon Detection *New Scientist* **138** (1868, 10 April) 20

Mullin T (ed) 1993 *The Nature of Chaos* (Oxford: Clarendon)

Nejoh H 1991 Incremental Charging of a Molecule at Room Temperature Using the Scanning Tunnelling Microscope *Nature* **353** 640–2

Nelson P G 1990 How Do Electrons Get Across Nodes? *Journal of Chemical Education* **67** 643–7

——1993 How do Electrons Get Across Nodes? *Journal of Chemical Education* **70** 346

———1994 Classifying Substances by Electrical Character *Journal of Chemical Education* **71** (1) 24–6

New Scientist 1991 Synchrotron Reveals Structure of Superconducting Fullerene **130** (1775, 29 June) 16

———1993 Big Bangs and Exploding Cakes **138** (1872, 8 May) 3

Norrby L J 1991 Why Is Mercury Liquid? *Journal of Chemical Education* **68** 110–3

Olson A J and Goodsell D S 1992 Visualizing Biological Molecules *Scientific American* **267** (5) 44–51

Orgel L E 1992 Molecular Replication *Nature* **358** 203–9

Orszag S A and Zabusky N J 1993 High-Performance Computing and Physics *Physics Today* **46** (3) 22–3 and rest of this special issue

Osada Y and Ross-Murphy S B 1993 Intelligent Gels *Scientific American* **268** (5) 42–7

Ottino J M 1989 The Mixing of Fluids *Scientific American* **260** (1) 40–9

Ottino J M, Muzzio F J, Tjahjadi M, Fianjione J G, Jana S C and Kusch H A 1992 Chaos Symmetry and Self-Similarity: Exploiting Order and Disorder in Mixing Processes *Science* **257** 754–60

Ouyang Q and Swinney H L 1991 Transition from a Uniform State to Hexagonal and Striped Turing Patterns *Nature* **352** 610–2

Overney R and Meyer E 1993 Tribological Investigations Using Friction Force Microscopy *MRS Bulletin* **19** (5) 26–34

Owers-Bradley J 1993 Science Goes on Below Absolute Zero *Physics World* **6** (7) 24–5

Pegrum C 1992 Natural Josephson Junctions *Nature* **358** 193

Pendry J B 1991 Photonic Insulators *Nature* **354** 435–6

Pentecost A 1991 Springs that Turn Life to Stone *New Scientist* **132** (1800/1801, 21/28 December) 42–4

Pepper D M, Feinberg J and Kukhtarev N V 1990 The Photorefractive Effect *Scientific American* **263** (4) 34–40

Petford N 1991 A Sticky Subject: Viscosity and Earth Materials *New Scientist* **130** (1773, 15 June) 41 within article 'Granite on the Move' 38–42

Phillips W D and Metcalf H J 1987 Cooling and Trapping Atoms *Scientific American* **256** (3) 36–42

Physics Today 1993 The Geometric Phase Shows up in Chemical Reactions **46** (3) 17–9

Pippard A B 1966 *Elements of Classical Thermodynamics for Advanced Students of Physics* (Cambridge: Cambridge University Press)

Pobell F 1993 Solid-State Physics at Microkelvin Temperatures: is Anything Left to Learn? *Physics Today* **46** (1) 34–40

Pohl F 1993 Nothing but Atoms and the Void *New Scientist* **138** (1872, 8 May) 41

Pollack G L 1991 Why Gases Dissolve in Liquids *Science* **251** 1323–30

Postle D 1976 *Fabric of the Universe* (London: Macmillan)

Prigogine I 1980 *From Being to Becoming* (San Francisco, CA: Freeman)

Protheroe P 1993 Are Picture Books Harmful? *New Scientist* **138** (1878, 19 June) 44–5

Quate C F 1991 Switch to Atom Control *Nature* **352** 571

Rae A I M 1986 *Quantum Physics: Illusion or Reality?* (Cambridge: Cambridge University Press)

Raimi R A 1969 The Peculiar Distribution of First Digits *Scientific American* **221** (6) 109–20

Raveau B 1992 Defects and Superconductivity in Layered Cuprates *Physics Today* **45** (10) 53–8

Reed M A 1993 Quantum Dots *Scientific American* **268** (1) 98–103

Rees G and Robinson B 1991 Designer Solvents for Clever Chemistry *New Scientist* **130** (1770, 28 May) 39–43

Riordan M 1992 The Discovery of Quarks *Science* **256** 1287–92

Robbins M O, Thompson P A and Grest G S 1993 Simulations of Nanometer-Thick Lubricating Films *MRS Bulletin* **19** (5) 45–9

Robinson W R 1992 *Chemistry: Concepts and Models* (Lexington, MA: Heath)

Rodriguez J A and Goodman D W 1992 The Nature of the Metal–Metal Bonds in Bimetallic Surfaces *Science* **257** 897–903

Rouvray D 1994 Elementary My Dear Mendeleyev *New Scientist* **141** (1912, 12 February) 36–8

Rowell J 1991 High Temperature Superconductivity (introduction to group of articles by Sleight, Jorgenson and Batlogg) *Physics Today* **44** (6) 22–3

Ruelle D 1991 *Chance and Chaos* (Princeton, NJ: Princeton University Press) (This book is reviewed in *Nature* **358** 2013)

——1994 Where Can One Hope to Profitably Apply the Ideas of Chaos? *Physics Today* **47** (7) 24–30

Rugar D and Hansma P 1990 Atomic Force Microscopy *Physics Today* **43** (10) 23–30

Ryder L 1992 The Standard Model *Physics Education* **27** 66–75

Salmeron M B 1993 Use of the Atomic Force Microscope to Study Mechanical Properties of Lubricant Layers *MRS Bulletin* **19** (5) 20–5

Sato A and Tsukamoto Y 1993 Nanometre-Scale Recording and Erasing with the Scanning Tunnelling Microscope *Nature* **363** 431–2

Schattschneider D 1994 Escher's Metaphors *Scientific American* **271** (5) 48–53

Scher H, Shlesinger M F and Bendler J T 1991 Time-Scale Invariance in Transport and Relaxation *Physics Today* **44** (1) 26–34

Scientific American 1967 *Materials* (San Francisco: Freeman)

Seddon J and Templer R 1991 Liquid Crystals and the Living Cell *New Scientist* **130** (1769, 18 May) 41–5

Selinger B 1989 *Chemistry in the Marketplace* 4th edn (Sydney: Harcourt Brace Jovanovich)

Shafer N E and Zare R N 1991 Through a Beer Glass Darkly *Physics Today* **44** (10) 48–52

Sheldrake R 1987 Mind Memory and Archetype: I Morphic Resonance and the Collective Unconscious *Psychological Perspectives* **18** (1) 9–25

Shimony A 1988 The Reality of the Quantum World *Scientific American* **258** (1) 36–43

Shinbrot T, Grebogi C, Ott E and Yorke J A 1993 Using Small Perturbations to Control Chaos *Nature* **363** 411–7

Silver J 1989 Chemical Chameleons for Electronics *New Scientist* **123** (1684, 30 September) 32–5

Sleight A W 1991 Synthesis of Oxide Superconductors *Physics Today* **44** (6) 24–30

Smalley R E 1991 Doping the Fullerenes *Fullerenes: Synthesis Properties and Chemistry of Large Carbon Clusters* ed G S Hammond and V J Kuck (*ACS Symposium Series* **481**) pp 141–59

Smith P A S 1992 Trivial Names for Chemical Substances *Journal of Chemical Education* **69** 877–8

Sorensen A H and Uggerhoj 1989 The Channeling of Electrons and Protons *Scientific American* **260** (6) 70–7

Sparberg E B 1996 Hindsight and the History of Chemistry *Journal of Chemical Education* **73** 199–202

Stamp P C E 1996 Tunnelling Secrets Extracted *Nature* **383** 125–7

Stapp H P 1982 Mind Matter and Quantum Mechanics *Foundations of Physics* **2** 363–99

Stein B P 1994 Atoms Caught in a Web of Light *New Scientist* **141** (1910, 29 January) 32–7

Stein D L 1989 Spin Glasses *Scientific American* **261** (1) 36–42

Stephens P W and Goldman A I 1991 The Structure of Quasicrystals *Scientific American* **264** (4) 24–31

Stephens P W, Mihaly L, Lee P L, Whetten R L, Huang S-M, Kaner R, Deiderich F and Holczer K 1991 Structure of Single-Phase Superconducting K_3C_{60} *Nature* **351** 632–4

Stevenson T 1996 Invasion of the Nanomachines *New Scientist* **149** (2014, 27 January) 48

Stewart I 1992 Where Do Nature's Patterns Come From? *New Scientist* **135** (1835, 22 August) 14

———1993 The Half-Life of a Dirty Book *New Scientist* **138** (1872, 8 May) 12

Stix G 1996 Trends in Nanotechnology: Waiting for Breakthroughs *Scientific American* **274** (4) 78–83

Sutton C 1993 World of Quarks *New Scientist* **139** (1881, 10 July) Inside Science No 63

Suzuki S, Green P G, Bumgamer R E, Dasgupta S, Goddard W A III and Blake G A 1992 Benzene Forms Hydrogen Bonds with Water *Science* **257** 942–945; also cover of this issue and Perspective on p 887

Tabor D 1969 *Gases Liquids and Solids* (Harmondsworth: Penguin, Library of Physical Sciences)

Tagg S L, LeMaster C L, LeMaster C B and McQuarrie D A 1994 Study of the Chaotic Behaviour of a Damped and Driven Oscillator in an Asymmetric Double-Well Potential *Journal of Chemical Education* **71** 363–74

Tanaka T 1981 Gels *Scientific American* **244** (1) 110–23

Taylor R and Walton D R M 1993 The Chemistry of Fullerenes *Nature* **363** 685–93

Templer R and Attard G 1991 The Fourth State of Matter *New Scientist* **130** (1767, 4 May) 21–5

Thall E 1996 When Drug Molecules Look in the Mirror *Journal of Chemical Education* **73** 481–4

Thomas L, Lionti F, Ballou R, Gatteschi D, Sessoli R and Barbara B 1996 Macroscopic Quantum Tunnelling of Magnetization in a Single Crystal of Nanomagnets *Nature* **383** 145–7

Tilley D R and Tilley J 1990 *Superfluidity and Superconductivity* 3rd edn (Bristol: Adam Hilger)

Tomalia D A 1995 Dendrimer Molecules *Scientific American* **272** (5) 42–8

Traub J F and Woznikowski H 1994 Breaking Intractability *Scientific American* **270** (1) 102–7

Uemura Y J, Keren A, Le L P, Luke G M, Sternlieb B J, Wu W D, Brewer J H, Whetten R L, Huang S M, Lin S, Kaner R B, Diedrich F, Donovan S, Gruner G and Holczer K 1991 Magnetic Field Penetration Depth in K_3C_{60} Measured by Muon Spin Relaxation *Nature* **352** 605–7

Umbanhowar P B, Melo F and Swinney H L 1996 Localized Excitations in a Vertically Vibrated Granular Layer *Nature* **382** 793–6

Urry D W 1996 Elastic Biomolecular Machines *Scientific American* **272** (1) 44–9

Uzer T, Farrelly D, Milligen J A, Raines P E and Skelton J P 1991 Celestial Mechanics on a Microscopic Scale *Science* **253** 42–8 van Haaren J 1996 Liquid Crystals Tilted by Light *Nature* **381** 190–1

van Oss C J, Chaudhury M K and Good R J 1988 Interfacial Lifshitz–van der Waals and Polar Interactions in Macroscopic Systems *Chemical Reviews* **88** 927–41

Vaughan J 1994 Phlogiston and Fire-Air *ChemNZ* (56) 6–11

Volkmuth W D and Austin R H 1992 DNA Electrophoresis in Microlithographic Arrays *Nature* **358** 600–2

von Baeyer H C 1993 *Taming the Atom* (Viking) (Reviewed in *New Scientist* **138** (1872, 8 May) 41)

Vos W L, Finger L W, Hemley R J, Hu J Z, Mao H K and Schouten J A 1992 A High-Pressure van der Waals Compound in Solid Nitrogen–Helium Mixtures *Nature* **358** 46–8

Walker J S and Vause C A 1987 Reappearing Phases *Scientific American* **256** (5) 90–7

Walton A J 1983 *Three Phases of Matter* (Oxford: Clarendon), and 2nd edn

Washburn S 1992 Trapped in Mid-Air *Nature* **358** 537–8

Watson A 1991 The Perplexing Puzzle Posed by a Pile of Apples *New Scientist* **132** (1799, 14 December) 15

Weber R L 1992 *Science with a Smile* (Bristol: IOP Publishing), including p 22 'Blake and Fractals'; p 89 'The Ergodic Skeleton in the Cupboard of Thermodynamics'; p 280 'The Certainty of Uncertainty'; p 349 'Bizarre Measures'

Welland M 1994 New Tunnels to the Surface *Physics World* **7** (3) 32–6

Wentland S H 1994 A New Approach to Teaching Organic Chemical Mechanisms *Journal of Chemical Education* **71** 3–8

Whitesides G M 1995 Self-Assembling Materials *Scientific American* **273** (3) 114–7

Whittle M and Bullough W A 1992 The Structure of Smart Fluids *Nature* **358** 373

Wickramsinghe H K 1989 Scanned-Probe Microscopes *Scientific American* **261** (4) 74

Wilkinson P B, Fromhold T M, Eaves L, Sheard F W, Miura N and Takamasu T 1996 *Nature* **380** 608–10

Wilson K G 1979 Problems in Physics with Many Scales of Length *Scientific American* **241** (2) 140–57

Winfree A T 1991 Crystals from Dreams *Nature* **352** 568–9

Witten T A 1990 Structured Fluids *Physics Today* **43** (7) 21

Wolsky A M, Giese R F and Daniels E J 1989 The New Superconductors: Prospects for Application *Scientific American* **260** (2) 44–52

Wolynes P G 1996 A Fresh Glass of Frozen Chaos *Nature* **382** 495–6

Wynn C M 1992 Does Theory Ever Become Fact? *Journal of Chemical Education* **69** 741

Yam P 1991 There's the Rub *Scientific American* **264** (6) 14

Zettl A 1993 Making Waves with Electrons *Nature* **363** 496–7
Zewail A H 1990 The Birth of Molecules *Scientific American* **263** (6) 40–6
Zhou O, Fischer J E, Coustel N, Kycia S, Zhu Q, McGhie A R, Romanow W J, McCauley J P Jr, Smith A B III and Cox D E 1991 Structure and Bonding in Alkali-Metal-Doped C_{60} *Nature* **351** 462–4

INDEX

(Section titles are in bold type and names are in italics)